T0255504

CAMBRIDGE LIBRARY COLLECTION

Books of enduring scholarly value

Earth Sciences

In the nineteenth century, geology emerged as a distinct academic discipline. It pointed the way towards the theory of evolution, as scientists including Gideon Mantell, Adam Sedgwick, Charles Lyell and Roderick Murchison began to use the evidence of minerals, rock formations and fossils to demonstrate that the earth was older by millions of years than the conventional, Bible-based wisdom had supposed. They argued convincingly that the climate, flora and fauna of the distant past could be deduced from geological evidence. Volcanic activity, the formation of mountains, and the action of glaciers and rivers, tides and ocean currents also became better understood. This series includes landmark publications by pioneers of the modern earth sciences, who advanced the scientific understanding of our planet and the processes by which it is constantly re-shaped.

A Catalogue of the Collection of Cambrian and Silurian Fossils contained in the Geological Museum of the University of Cambridge

John William Salter (1820–1869) was an English naturalist and geologist, best known for his work as palaeontologist to the Geological Survey of Great Britain. This is a complete catalogue of the Cambrian and Silurian fossils in the Geological Museum at the University of Cambridge. Preceded by a detailed introductory section on the Palæozoic system, the catalogue is arranged by geological strata, covering the various groups of Cambrian and Silurian fossils. The entries include detailed illustrations, along with references to the location of each fossil in the collection, its name and details of its place of origin. Revised by staff of the University and published posthumously in 1873, the catalogue also contains a substantial preface by Adam Sedgwick, famous for his role in the development of modern geology,which provides fascinating insights into the geological advances of the Victorian era.

A Catalogue of the Collection of Cambrian and Silurian Fossils contained in the Geological Museum of the University of Cambridge

J.W. Salter
Adam Sedgwick
John Morris

CAMBRIDGE UNIVERSITY PRESS

Cambridge, New York, Melbourne, Madrid, Cape Town, Singapore,
São Paolo, Delhi, Dubai, Tokyo, Mexico City

Published in the United States of America by Cambridge University Press, New York

www.cambridge.org
Information on this title: www.cambridge.org/9781108015943

This edition first published 1873
This digitally printed version 2010

ISBN 978-1-108-01594-3 Paperback

A CATALOGUE

OF THE

COLLECTION OF CAMBRIAN AND SILURIAN FOSSILS

IN THE

GEOLOGICAL MUSEUM OF THE UNIVERSITY OF CAMBRIDGE.

London:
CAMBRIDGE WAREHOUSE, 17 PATERNOSTER ROW.
Cambridge:
DEIGHTON, BELL AND CO.

THE REV. ADAM SEDGWICK, LL.D., F.R.S.,

WOODWARDIAN PROFESSOR OF GEOLOGY, AND SENIOR FELLOW OF

TRINITY COLLEGE IN THE UNIVERSITY OF CAMBRIDGE.

From a picture by Lowes Dickinson, belonging to A. A. Van Sittart, Esq.

A CATALOGUE

OF THE COLLECTION OF

CAMBRIAN AND SILURIAN FOSSILS

CONTAINED IN

THE GEOLOGICAL MUSEUM OF THE UNIVERSITY OF CAMBRIDGE,

BY

J. W. SALTER, F. G. S.

WITH A PREFACE BY

THE REV. ADAM SEDGWICK, LL.D. F.R.S.
WOODWARDIAN PROFESSOR OF GEOLOGY IN THE UNIVERSITY OF CAMBRIDGE,

AND

A TABLE OF GENERA AND INDEX

ADDED BY

PROFESSOR MORRIS, F. G. S.

Cambridge:
AT THE UNIVERSITY PRESS.
M. DCCC. LXXIII.

Cambridge:
PRINTED BY C. J. CLAY, M.A.
AT THE UNIVERSITY PRESS.

CONTENTS.

PREFACE.

Since the proof-sheets of the following Catalogue issued from the press, I have earnestly desired to write an Introductory Preface to it. Week after week and month after month have I waited in anxious hope of completing my humble task. But I have been greatly interrupted by a chronic malady which makes me incapable of any long-continued mental labour; and, in addition to this hindrance, a painful infirmity of sight almost entirely prevents me from consulting my manuscripts and memoranda, made during the Geological tours of many past years. To spare this continued infirmity of sight I now gratefully dictate the following pages to my young friend and assistant in the Museum—Mr Walter Keeping.

The old Catalogue of the Palæozoic Fossils, by Prof. McCoy (now of the University of Melbourne), was a work of enormous labour and of very great scientific skill; especially when we consider the date of its appearance. Its publication was a real benefit to the Academic Student, a distinction to the University Press, and a great honour to its Author. In the clearness and elaborate accuracy of its descriptions of the several species the work is I think unrivalled, in spite of all that has been written since. But I wish to write historically, and profess not here to enter upon critical questions of scientific detail.

In an advancing science like Geology, any catalogue, however good at the time of its publication, must soon become defective from changes of nomenclature, from improved classifications, and above all from the discovery of new species. Mr Davidson's great works on the Brachiopoda have thrown new light upon the divisions of that class of Mollusca, and very greatly changed the nomenclature of the genera and species.

The great additions made, of late years, to our knowledge of the older Palæozoic Fauna, and especially to the groups of fossils now in our Museum, derived from the lower division of what were formerly called the Lingula Flags; still more the great Fletcher Collection from the Wenlock series, which has been purchased by the University since Prof. M[c]Coy's Catalogue was printed—all these additions prove the necessity of corresponding changes in our Palæozoic Catalogues.

After Mr Salter had left the service of the Government Survey he several times visited me at Cambridge, and was desirous to do his best to supply the imperfections of our Catalogue; and I joyfully accepted his offered services, knowing his great skill as a Natural Science Artist, and believing him, after his long-tried labours under the Government Survey, to be unrivalled in his exact scientific knowledge of the fossil Invertebrata of the British Isles. One condition of his engagement was that his work should be constructed as a Supplement to the old Catalogue, and not as an independent work. This condition Mr Salter professed to accept, though as his labours advanced he did not by any means perfectly conform to it.

To his personal applications, seldom communicated to myself, and to the knowledge that he was employed in completing a Cambridge Palæozoic Catalogue, we owe many of the additions to our older Palæozoic Collection, to which I have before alluded. And let me here record in behalf of the University my grateful thanks to the Earl of Ducie, to David Homfray, Esq. of Portmadoc, to Dr Hicks of St Davids, to Mr Lightbody of Ludlow, to Mr Ash, and to Mr J. Plant of the Royal Museum, Salford, for their generous donations to our Collection; and I trust that the several donors have found a grateful and respectful notice in Mr Salter's pages. If the names of any other benefactors have been here omitted, I can only plead in my behalf the death of Mr Salter before the Catalogue was quite complete, my present infirmity of sight which prevents me from consulting my Journals and Memoranda, and the strange clouds of oblivion which too often trouble an old man's memory.

Mr Salter's task advanced very slowly; for his bodily health and his nervous system seemed to have been almost broken up by the stress of hard mental labour which had been imposed upon him through many preceding years. On several occasions he abandoned his task at Cambridge, and went to recruit his health by a residence of a few months at Malvern, where he remained under medical care. But still the work did advance in spite of these interruptions, and the University at length undertook to pay the cost of it under certain conditions, with which he was willing to comply.

After many delays and much anxiety (and I may add after much personal cost to myself) Mr Salter's manuscript took a form which made it fit for the press. I do not however think it would ever have reached that state but for the kind advice and encouragement and judicious help given to its author by the Master of St Peter's College. And after the first press-work was done, it still was evident that the Catalogue required careful revision and correction; not in the naming and description of the species, but in such marks of reference as should intelligibly connect the Catalogue with the specimens arranged in the Cabinets of the Museum. Without this kind of labour the best Catalogue in the world would be of comparatively little use to a Cambridge student. Knowing this, I was induced to take a journey to Margate for the express purpose of seeing Mr Salter, and making arrangements for the discharge of all his personal expenses while he was employed in revising the references of his Catalogue, and giving the ultimate finish to his work. The state of his health was perfectly deplorable when I last saw him at Margate, and the hand of death very soon afterwards arrested all our hopes of obtaining the Author's final revision of his work.

Again, we were, in an hour of need, under a great obligation to Dr Cookson, who undertook the laborious task of revising and completing the exact agreement of the cabinet-labels with the details given in the several pages of the Catalogue. This work of Dr Cookson's not only required great labour, but very nice discrimination in the separation of the species, and demands from us the warmest expressions of our gratitude.

The final scientific revision of the work was undertaken, at the request of the University, by Prof. Morris, who went through his task with that genuine conscientious and laborious skill which left nothing to be desired: and it was well for the reputation of the University that this final labour was undertaken by so distinguished a Palæontologist; for there were several blemishes in the Catalogue: such, for example, as the appearance of the same species under different names, a few mistaken localities, and other little incongruities arising out of Mr Salter's wretched health and continually interrupted labours.

Since Prof. Morris' last revision and Report there has been again a long delay, for which I alone am responsible; and I trust that the University will accept this acknowledgment as an apology from their very old and infirm, but, in former years, their very hard-working servant.

The following Catalogue of all the Older Palæozoic Fossils in the Woodwardian Museum consists of specimens which have been collected by myself, or presented by my friends, or obtained by purchase during my very long tenure of office as Professor of Geology.

It was meant to be a Supplement to the Synopsis of Palæozoic Fossils by Prof. McCoy; but this condition, as before stated, was not strictly observed; for it will be seen, among other variations, that the nomenclature of Prof. McCoy has not been always followed, the subsequent examination of some of the specimens by Mr Salter having induced him to class them under other genera than those under which they appeared in McCoy's "Synopsis."

The Collection is, to a great extent, arranged zoologically, under the several Geological divisions of the Strata adopted in this Catalogue; and the fossils of each division, commencing with the lowest, are tabulated under four columns—the first column indicating the number of the Table-Case and column of Drawers (as Gh. 1); the second column giving illustrative figures of various genera, and also references to McCoy's Synopsis; the third giving the names adopted by Mr Salter, with references to one or more works in which they have been described or figured, with the addition of useful short notes on the species; the fourth column giving the localities from which the specimens have been obtained.

I should ill discharge my duty in writing this Preface did I not gratefully notice the elaborate Index which Prof. Morris has added to the Catalogue. It is a graceful finish to the work, and makes the Catalogue fit for ready consultation, which is a matter of the first importance. This Index has been a work of such labour that I should not have ventured to ask the Professor to undertake it; but this thought makes me the more grateful to him for having contributed it so kindly and spontaneously. With like expressions of grateful goodwill I must also mention the Tables, in which is given an account of the whole range of formations within which each genus has been found; so that an eye-glance at these Tables will put a Palæontologist in possession of the leading facts of the distribution of the organic types in the successive groups of Strata, as they are enumerated in the Catalogue: not in any hypothetical order, but in that in which they are recorded in Nature's Book. These Tables are a work of great knowledge and of patient skill; and went far beyond the task of revision intended by the University.

The stratigraphical System of nomenclature adopted in this Catalogue is essentially the same with that of Prof. McCoy's Synopsis. It is based upon an actual survey, first made by myself, whereby I approximately determined in N. Wales the order of the older deposits of the whole region, and the natural groups of strata into which they might be separated.

This might be called a great but rude problem of solid geometry, to be first solved by an elaborate examination of physical evidence, and without reference to the organic remains in the successive groups. But these groups being once established, on the basis of true observation, we may then proceed to obtain the first chapters of a true history of the succession of organic types, as the tale is told in the successive strata whence they have been derived. And when we have once obtained in any extensive section a true succession of organic types, we may then, as Nature is true to her own workmanship, advance a step farther, and use that succession to help us in making out the order of the Physical groups in cases where they have been imperfectly or obscurely elaborated. Thus we have two great principles of arrangement; first by the actual and laborious observation of the successive physical groups; secondly, by the order of the organic types which have been already established by a reference to the types of some well-known natural section.

In determining a Geological nomenclature these two great principles must never be lost sight of—No true nomenclature can be in conflict with the actual succession of the physical deposits; neither can it contradict the true succession of organic types. Nature does not contradict her own workmanship. This was the principle on which William Smith, whom we call the Father of English Geology, acted; and it was the principle on which Murchison acted when he first made known his beautiful succession in the upper part of (what he taught us to call) his Silurian System. That upper part of his System was thoroughly and beautifully worked out, was accepted at once, and continues to maintain its place. But below the Wenlock shale, in what he called the Lower Silurian groups, his fundamental sections utterly broke down, having no base to rest upon. He never made out the succession of his physical groups: some of them which required separation he confounded, and some he put in an inverted order; and thereby he brought an inevitable incongruity into his lists of the Older Palæozoic fossils. In short, I venture to affirm, that the Lower Silurian nomenclature, however widely adopted on the authority of its Author, was false: because it was built upon sections that were untrue to nature; and if this assertion be true—and it is true—the discussion requires no further argument.

As a general rule, honest truth and good taste go hand-in-hand; and what can be more incongruous and tasteless than to erase the Classical name of Cambrian as applied to the grand mountain chains of Caernarvonshire and Merionethshire, and to substitute the word Silurian as their designation. This was done by the Author of the "Silurian System," in the first instance no doubt by mistake, and in the hope of giving a greater extent and firmer basis to his System. But when the great errors of his fundamental groups were discovered, why continue such a monstrous abuse of nomenclature? Siluria supplies us neither with the best types of the older groups, nor with any sections which clearly define their succession: Cambria supplies both. Our business here is not to consider what great services the Author may have done in other regions: but to consider whether his work in Lower Siluria be true to nature. The first publication of his grand lists of "Lower Silurian Fossils" was a great boon to Geology; but the assumed stratigraphical arrangement and the grouping of the species has been a great mischief, and a drag upon its progress.

I will pursue this subject no farther, but refer the reader to my Introduction to Prof. McCoy's Synopsis which I here adopt, because that which is true in a natural arrangement can never be materially changed. Nor should I have introduced even this notice of an old controversy, had it not been revived in an acute, animated, and very elaborate dissertation by Prof. Sterry Hunt (*Nature*, May 2, 9, and 16, 1872, reprinted from the

Canadian Naturalist). And it was high time that a shameful incubus should be shaken off from the breathing organs of the older Palæozoic rocks; and that they should express themselves once again in the language of truth and freedom.

Tottering as I now am under the infirmities of old age, with faded senses and a failing memory, I am ill fitted for the part of a gladiator; but after Professor Sterry Hunt's bold and honest vindication of my work in Cambria, it would be on my part an act of moral cowardice, and a want of proper respect for the sanctity of evidence, not thus to speak out in the cause of historic truth and reason.

Leaving all controversy for awhile, I may mention some other facts respecting the Catalogue, which may find a proper notice in this Preface. I never saw the proof-sheets of the Catalogue or superintended its long progress through the press. There are consequently a few notices and expressions in it which do not strictly represent my views. The Sections on page 1, and page 3, were made and struck off without any communication with myself, and on Mr Salter's authority they entirely rest. Instead of the groups Harlech and Longmynd—the second from the base of the Ideal Section on page 1—I should have preferred to write Llanberis, Bangor, and Longmynd Groups, and I should have placed the Harlech Group at the base of the Menevian. The second Section, page 3, has I think considerable value; not as a natural Section, but as a pictorial illustration of the relative position of certain important groups of strata. The third plate, page 9, may be excellent in its way as a small Geological Map; but it is far too complicated for the use of a student unless it had been illustrated by sections.

I may here remark generally, that I do not think the Upper and Lower Llandovery Rocks of the Government Survey ought to appear as one group; and I would place the Lower Llandovery at the top of the Cambrian groups, and the Upper Llandovery at the base of the true Silurian Rocks. This is the arrangement justified by the Sections in Denbighshire: but there is a large extent of country, partly covered by the Upper and Lower Llandovery Rocks, with which I am very imperfectly acquainted, and which is, I think, even now, very inadequately described in the works of the Government Survey.

The separation between Cambrian and Silurian Rocks is sometimes defined by a simple line, which shows at once the discordancy of the two deposits. On the contrary, the passage between the two systems is not unusually marked by a great confusion of deposits, by enormous masses of stratified conglomerate violently contorted or set up on edge, which are certain indications of a vast period of time. Many monuments of powerful elevation, abrasion, and dislocation of the rock-masses also mark the long period of time occupied in the passage from the Cambrian to the Silurian formations.

I do not profess perfectly to understand, or to criticise, the comparative schemes of classification and nomenclature given by Mr Salter on page 25. I have a similar, more simple comparative Table, kindly given me by Prof. Morris, which I will not here copy *in extenso*, as I wish this prefatory notice to be short; but I will copy the first column of his Table, which he believes to represent my scheme of arrangement as well as his own; and with which I wish to make no change; except that I would add the Llanberis Group to the Longmynd and Bangor Groups, and remove the Harlech Group to the base of the Menevian, as before stated.

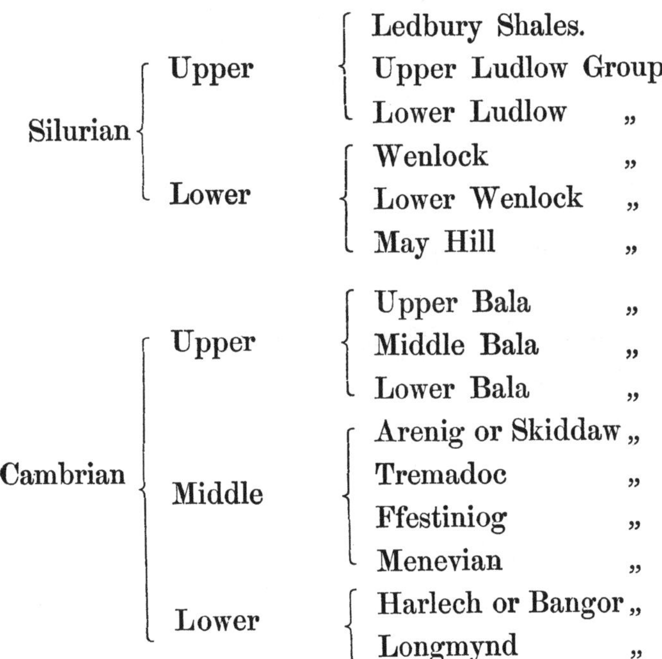

Silurian	Upper	Ledbury Shales.
		Upper Ludlow Group.
		Lower Ludlow "
	Lower	Wenlock "
		Lower Wenlock "
		May Hill "
Cambrian	Upper	Upper Bala "
		Middle Bala "
		Lower Bala "
	Middle	Arenig or Skiddaw "
		Tremadoc "
		Ffestiniog "
		Menevian "
	Lower	Harlech or Bangor "
		Longmynd "

If it be asked how the great succession of our Older Palæozoic Rocks was determined, I can only refer to my own individual labours, carried on during many successive years among the older rocks of England.

I commenced my task in Wales in 1831, accompanied for a short time by my friend Charles Darwin—a name now well known and honoured in the whole world of science—but other engagements soon drew him away from N. Wales. We laid down the northern boundary of the Carboniferous Limestone together, and I at first purposed to examine the rocks in descending order. But I found in Denbighshire the same interrupted broken masses of Old Red Sandstone which I had so many times noticed between the Carboniferous and the Silurian and Cambrian rocks in the North of England. I therefore despaired of establishing a good base to work upon in that district; and after a short examination of the rocks on both sides of the Menai Straits, I resolved to fix a provisional base-line on the Caernarvon side of the Straits.

I soon found that the prevailing strike of the country was about N.N.E.; therefore by following nearly East and West sectional lines, the deposits were discovered in their true relations and sectional order: and commencing nearly on the line of the Holyhead road, I marked all the anticlinal and synclinal flexures upon a section copied from nature. The vast undulations of the strata, the constant intrusion and alternation of igneous rocks, at first threw some difficulties in my way; but I had learned to encounter them during my previous three years' labour among the Lake mountains, and I did not find any insurmountable difficulties in reading the succession given by each section. In this way I made three parallel nearly East and West traverses. The first from the shores of the Menai, by the great Penrhyn slate-quarries, and so on, by the summits of the high mountains, over the top of Glyder Fawr to the neighbourhood of Capel Curig: the second, by the mountains South of Llanberis, over the top of Snowdon, and thence into the valley above Beddgelert: the third, by a parallel mountain-track, passing over Moel Hebog and ending in the valley below Beddgelert, a little farther to the South-west.

On a careful revision of the several sections presented on these three lines, I found that even in the most complicated curves the chief anticlinal and synclinal lines might be brought into an approximately close comparison, so as greatly to assist in the construction of a general section of the country and the establishment of good physical groups. Thus the great synclinal trough of Cwm Idwal, on the first line of section; another synclinal in the second line, within which rests the top of Snowdon; and a third synclinal which underlies the top of Moel Hebog: all affect one group of strata, through which pass the most remarkable fossiliferous beds of the whole Cambrian Series.

Taking this as a kind of key-note to guide me in further discoveries I examined the sections in more detail, and made out the relations of their corresponding parts; and before the working season was over (in 1831) I completed an approximate Geological Map from actual survey of nearly the whole of Caernarvonshire. No names were of course given to the natural groups of strata; but the part some of them played in the physical development of the country was, I thought, clearly established.

I have given these details to shew the honest and very laborious way in which I set about my work. I had, in truth, little difficulty in reading the sections, as their language was written in characters almost identical with those I had long studied in the Cumbrian mountains.

The base-line on the shores of the Menai was broken and imperfect; and its association, made in after years by the Government Survey and by myself, with the Longmynd rocks, was, I think, partly hypothetical. But along each line of section, above described, I had found a magnificent, and, on the whole, an ascending series of deposits, of grand features and of enormous thickness.

Early next summer I resumed my task in Wales. Crossing the depression in which my sections had terminated, I pushed them forward towards the East, nearly at right angles to the strike of the strata; and in the progress of my work, discovered the great Merioneth-Anticlinal, which I have often called the backbone of Wales. It brings out the oldest strata which were first seen near the coast of the Menai; and assuming it as our base, we can count off towards the East an enormous series of ascending strata capped by the whole Bala group; and from the same base-line we can count off all the groups of the three sections observed in the preceding year, first in ascending, and then in descending order, till we meet the same basic groups as we approach the shores of the Menai.

Having thus obtained a key-note to the harmonious grouping of the strata, and having practically become acquainted with some of the most important physical groups, I undertook what proved to be the severest summer's task of my Geological life; namely, the interpretation and partial delineation of the order and principal flexures of all the older deposits of the counties of Merioneth, Montgomery, and Denbigh.

A brief synopsis, illustrated by sections of what I had effected in Caernarvonshire, was laid before the British Association, at one of their evening meetings held at Oxford in the year 1832. I had no doubt about the great groups or about the great flexures and faults by which some of them were repeated again and again in the same county.

I never had a Geological secret in my long life. Nearly all my best work in Wales was done in solitude, and was therefore my own. My first groups continue unchanged and unmodified; with the exception of certain changes introduced by recent discoveries, such as the Menevian group, which now forms a group subordinate to the Lingula Flags; and I hesitate not to affirm that the grand and well-connected succession of deposits which I unfolded between the Menai and the top of the Berwyn chain is unrivalled by any other European Section, of the same age, hitherto described by Geologists. The Cambrian sections have this crowning honour; and are rivalled in their succession and physical development only by the magnificent series of Palæozoic rocks discovered by the Geologists of North America.

What sense therefore has there been in excluding the Mountains of Wales from their proper physical importance in the Geology of our own Island by sinking them and colouring them as Lower Silurian? The groups of the Lower Silurian System of Sir R. I. Murchison, (even had their place and age not been utterly mistaken by their Author,) would not have deserved the prominent notice they have held in the nomenclature of English groups; for they generally want the essential condition of good typical groups. They do not shew any true relation to the groups above them and below them.

For my own convenience I had made an agreement with the Author of the Silurian System, at the Edinburgh Meeting of the British Association, in 1834, to wait till he had finished the result of his labours, which I expected to appear in a well-illustrated volume in course of the year following. But having waited, as I thought, too long for the appearance of the "Silurian System," I gave up my sketches, day-books, and field-books (I think in 1837) to my honoured friend Mr Lonsdale, then the Assistant-Secretary of the Geological Society, and out of these documents he made a series of sections upon a grand scale, which shewed the extent, and many of the details, of my work in North Wales. These sections were exhibited and explained by myself to the Geological Society at one of their evening meetings, and remained in their possession for many years. The last time I saw them they were in the hands of Mr Warburton, who had undertaken to reduce to a state fit for publication the papers on North Wales, which described the labours of Mr Salter and myself in the years 1842—3, and he had obtained Mr Lonsdale's larger sections for assistance in this work.

In order to give coherence to my scattered remarks on the older rocks of North Wales, I will first mention in chronological order the chief periods during which I investigated the structure of the Principality. My best work, I think, was done in the summers of 1831 and 1832, in the manner above stated. In the spring and summer of 1833 my health broke down so much that I was incapable of taking the field till the autumnal season, when it was far too late for me to attempt the great and difficult task I had proposed to myself—namely, of commencing with the South flank of Cader Idris, and thence, by numerous long traverses, connecting my work in North Wales with the typical Silurian country on the banks of the Towey*.

After studying the Sections which were laid by Murchison before the British Association in 1833, I felt convinced that there was an overlap between the Systems of Cambria and Siluria (as they were afterwards called), and we agreed to settle this question next year by a joint tour through the most typical portions of the Silurian country, which had been, during the preceding years, examined, mapped, and described in considerable detail. By this joint labour we hoped to clear up some points of difficulty, and to establish a good line of demarcation between our Groups of Strata.

We commenced our work (in 1834) by various hasty traverses in the typical Silurian country, which stretches on both banks of the Towey. At first we had no matter of controversy, for I accepted at once my friend's interpretation of his own Sections. I did not go to dispute, but to learn as it were the alphabet of the Silurian tongue. My

* The autumn of that year was however not without its fruit: for accompanied and assisted by the present Astronomer Royal and Dr Whewell, afterwards Master of Trinity College, I made out in considerable detail the structure of Charnwood Forest, and determined the range of its single anticlinal axis; in following which towards the North we found that it brought up at a high angle of elevation two singular masses of dolomitized Carboniferous Limestone; but out of the line of disturbance the Limestone regained its ordinary type.

friend had a Government Map of the Country we first glanced at, well marked and coloured, with references to his previous labours. I had been disappointed in my hope of procuring the map of the Geographical Survey as I passed through London, and I had no map whatever with me of that part of the country; but I gradually learnt, on the spot, to understand what were the characters of such groups of Strata as were afterwards known to all British Geologists by the names Caradoc and Llandeilo rocks.

As we advanced northwards and passed the limits of the published Ordnance map, I could then turn to good account my old field map; and when I found myself among the rocks I had hastily examined in the year 1832, I readily accepted my friend's interpretation of the calcareous beds of Meifod, of the calcareous bands among the undulating rocks between Meifod and Llanfyllin, and the still more astonishing groups which are displayed between the Tannat and the Ceiriog. In these districts all the calcareous bands were counted as Caradoc. But my scepticism was alarmed before we reached the Ceiriog; because the calcareous bands, which descend from the northern end of the Berwyn chain into the valley of the Ceiriog—ranging about E.S.E., and nearly at right angles to the strike of the northern Berwyns—appeared to me, on almost certain evidence, to be only branches given off from Bala limestone. Here was a great difficulty. For if the Glyn Ceiriog limestone could not be separated from the calcareous bands we had left behind us, and if they were Caradoc, this limestone must be also Caradoc: and the natural conclusion drawn from my Sections was, that the Caradoc sandstone was exactly on the same parallel with the Bala limestone. But how could this be reconciled with the Silurian Sections of Murchison? For on his scheme, both the Caradoc and the Llandeilo groups were several thousand feet above the Bala limestone and its associated calcareous slates. To settle this difficulty we retraced our steps to Llangynog; and from thence we crossed the Berwyn Chain, marking its structure and the synclinal position of the Bala limestone by the way: and so we descended to some highly fossiliferous quarries in the limestone, which I had examined two years before.

There we parted, never to meet again in North Wales. I gave my friend all the chief localities of the Bala limestone, in its long range towards the South-Eastern flank of the Cader Idris group.

He did follow the line of the Bala limestone; but he gave me no information respecting his labours or his discoveries; nor did he tell me then, nor did I ever know, before the publication of the "Silurian System," that every species of the Bala Limestone fossils, which we collected together, and all of which he carried away for examination, were well-known Caradoc species.

We met, however, again after the end of the same summer, at Edinburgh, during the meeting of the British Association, and I naturally enquired what had been the success of my friend's excursion along the strike of the Bala Limestone, and whether he had seen reason to make any change of position among his typical groups. He replied that he had followed the Limestone as far as the South-East flank of Cader Idris; and that he had no change to make in the position and relations of his fundamental groups (the Caradoc and Llandeilo); in short, that after careful re-examination of his Sections and Classifications, he had no mistake to correct.

I had nothing to oppose to this, except my great surprise; and after reconsidering, for a while, my own Sections, I informed him, at the time of our discussion, that I could think of no place for his lower Silurian groups, unless they had disappeared among the rocks at the northern end of Berwyns, along the line of unconformable junction; either by an overlap of the upper or true Silurian rocks of Denbighshire, or, by some mistake of mine, in appreciating the calcareous and fossil-bearing groups, I had seen along that line: and to illustrate my view, I drew a rough outline sketch to shew how it was possible to interpolate, at the Northern end of the Berwyn and Glyn Ceiriog range, some beds, which might be higher than the Bala group, and on the parallel of the Caradoc sandstone. But I did this not with any view that my rough sketch should be published: for it was but an artifice to escape from a very great difficulty. It was in itself improbable, and it was based upon the *assumption that the Lower Silurian Sections were true.* Assuming this as a fact, I suggested, as a mere hypothesis, that the two groups, Caradoc and Llandeilo, might be concealed, or perhaps obscurely represented along the great line of discordant junction, which separated my Cambrian from his true Silurian rocks, in Denbighshire.

Such is the history of that Section which appears with my name affixed to it, at the North-East end of the great map of the "Silurian System." The author of that map had no authority from me to publish the hypothetical sketch. It may appear strange that I should think it worth while thus to dwell upon a minute point which is now seldom seen or thought of; but it was, in fact, the very pivot on which my dispute with Sir R. I. Murchison turned.

After we parted at the rich fossil-quarry near Bala, we never had one single syllable of correspondence respecting the older Palæozoic rocks of North Wales; nor did I again explore a single quarry of those rocks till full eight years had passed away. But, as before stated, the great work, the "Silurian System," appeared in the early summer of 1839; and when I saw my hypothetical Section entered upon the map, with my name affixed to it, I thought I had some right to be offended at this liberty: for the author had invited me to colour that part of his map which was east of the Berwyn chain; and this I refused to do, on the ground of my incompetency, at that time, to colour the country correctly; and I should have given unquestionably a like refusal, if he had alluded to the sketch above described. My mistake was in believing, on the authority of their author, that the lower Silurian Sections were correct. I did not, however, mean to let the matter rest, but to re-examine the whole Palæozoic question as soon after the publication of the Silurian System as I could find a long vacation at my command.

After I had parted with my friend near Bala in 1834, I thought the rest of the summer well employed in making traverses through the true Silurian rocks of Denbighshire, and in partially exploring the Carboniferous rocks of Denbighshire and Flintshire. In 1835 I spent some time in exploring the North of Ireland, after the breaking-up of the British Association at Dublin. It had been confidently asserted during the Meeting, that our Geological theories were put to open shame by the Culm-Measures of North Devon, which (though containing beds of coal and many true Carboniferous fossil plants) were in fact interpolated among the oldest Slate rocks; and that the Geological Society were discredited in not having given a proper prominence to this notorious fact.

In these expressions of vituperation, a challenge seemed directed personally to Murchison and myself, which we accepted with perfect goodwill. Accordingly, we visited the Northern Coast of North Devon next year, and resolved, if possible, to determine what was the true position of the Culm-Measures. We had no difficulty in making out the Natural groups of Strata, which presented themselves in a traverse from the extreme Northern Coast to the dark coloured limestone, which ranges a little South of Barnstaple, and crosses a great part of the County in a direction nearly East and West. Our attempt was at length quite successful. The Culm-Measures were proved to be the highest rocks of North Devon, and though anomalous in many of their mineralogical details, they were by no means anomalous in their position; for the calcareous bands which appeared along their northern and their southern base, were but one of the forms of the Carboniferous limestone.

But what were the groups of Slate rocks, which rose from beneath the Culm-Measures, and were exhibited in various undulations, ranging nearly East and West between the parallel of Barnstaple and the northern shores of Devon? The fossils in these groups exhibited some forms that were then unknown to us. But the highest group of all, which we will call the Barnstaple group, was eminently fossiliferous, and was pronounced by Murchison to belong to the Caradoc sandstone. We collected from this group a good series of organic remains, and sent them to Mr Sowerby; and not long afterwards my fellow-labourer received a dispatch from Mr Sowerby, which informed him that he had made a correct determination of the age of the Barnstaple fossils.

I was little satisfied with this determination, which virtually cost me the work of two Long Vacations in the years 1837 and 1838. In 1837 I was joined for a short time in South Devon by my friend Mr Godwin-Austen, and we did not quite complete any part of our survey, both being unexpectedly called away. But during a long Summer of 1838 I worked almost in solitude. I completed my traverses in North Devon and South Devon, and I traced Fossil-bearing Strata on the South-eastern Coast of Cornwall; and then doubling round from Penzance to the Northern Coast of the County, I obtained fossils partly by digging them from the rocks, and partly as gifts from my friends; and continued my way till I found myself once again among the rich Fossil-bearing quarries of Petherwin.

Late in the autumn of 1838 I brought back with me good sectional and palæontological evidence, which seemed to prove that nearly all the groups in the two Counties I had examined were of an older date than the Carboniferous rocks of North Devon, and of a newer date than the newest rocks in the system of Siluria. When I expounded this evidence to Murchison, he opposed it by a succession of ingenious hypotheses, which could not however stand against the simple evidence of my Sections. But to settle this point for ever, I proposed that we should adjourn to the house of Mr Sowerby, and, if possible, re-examine the hamper of fossils we had sent to him in the year 1836. The hamper was found in the exact state in which we had last seen it; nor do I believe that Mr Sowerby had ever opened it. However that might be; on opening the hamper, we saw a very good series of Devon Fossils with well-marked localities; but we saw nothing resembling a characteristic Caradoc species. Thus the Devonian System gradually became established, and the results from Sectional and Fossil evidence were in perfect harmony; and thus we took the first step, which I followed up in subsequent years; and much good work has been done since among the rocks of that series.

The next year formed an epoch in the history of European Geology; for in the early part of 1839 the "Silurian System" was first published. It was beautifully embellished and contained an accurate delineation and description of the most ancient Palæozoic Fossils of a large portion of Wales and some of the adjoining counties, such as had never before appeared in any Geological work. For it professed to arrange the lists of Fossils and the Groups of Strata in a true order of superposition. It had cost the Author seven years' field labour, and he was assisted by three distinguished naturalists in determining the classification and the nomenclature of his multitudinous fossils. It is no part of my duty to attempt a task far beyond my power—viz. to assign the proportional honours due to each of the scientific workmen who had contributed to the great work. But the chief honour will ever be given to the author of the System, who brought the materials together and arranged them in that manner in which they are seen in his splendid work. Under his hands the older Palæozoic Geology had assumed a new and a nobler type, and the highest praise was given to his work in all the scientific Journals of Europe and the United States; and as years advanced new honours accumulated on the author's head.

During the summer of the same year (1839), I joined my friend in a visit to the Rhenish Provinces and the North of Germany, for the purpose of following out those conclusions respecting the Devonian System which we had arrived at the preceding year, chiefly through my personal labours. Our summer work suffered a considerable retardation from a premature attempt (sanctioned by no less a personage than Prof. A. Goldfuss of Bonn) to classify the fossils of the Eifel Limestone with those of the Upper Silurian Groups; but before the expiration of the summer we escaped from this difficulty; and as our joint labour has been published, it would be idle for me to dwell upon it any longer in this Preface.

For nearly three months during the Academic vacation of 1840, I was confined by my duties in the Cathedral of Norwich, and was therefore cut off from any extensive field work: but within the limits of the vacation, guided by the Silurian map, I made some hasty excursions which brought to the Cambridge Museum what I then regarded as a rich harvest of fossils.

The year 1841 was partly employed by me in studying the Devonian rocks of Ireland under the guidance of my friend Sir Richard Griffith; and from Ireland I passed into Scotland, still in quest of facts that might give me the means of constructing a classification which would apply to every portion of the older rocks of Great Britain; but neither during that nor during any other tour did I find anything to compare with, much less to supersede, the magnificent succession of groups which I had seen in Wales and Siluria.

I resolved therefore to re-examine the whole of my work in Wales, and then to perform the same task among the Lake mountains and the districts bordering upon them. In this way I endeavoured to bring the several Palæozoic Groups into good co-ordination, and to name the lower portion of them in conformity with the system of Cambria, and the upper portion in like manner in conformity with the system of Siluria. This task I hoped to complete in two hardworking summers; but I found to my cost that I had greatly underrated the labour that was before me.

In 1842, with Mr Sowerby's permission, I was joined by Mr Salter, as a youthful and then joyous fellow-labourer, and especially as one well prepared, to complete the fossil catalogues of the several groups which we had to examine. This task employed us during two entire hardworking summers.

At the end of the summer of 1843, we had done our work thoroughly, as I then thought; for my previous labours in North Wales enabled me to conduct my young friend and assistant to all the localities of principal interest, with small loss of time in seeking them out. For a few days in 1842, we were joined by Sir Richard Griffith; and I will here state, in as few words as I can, the results of this joint work of two summers. My youthful and cheerful companion gradually became a good field surveyor, and he dressed up my Sections so as to make them fit for publication, of course on a reduced scale, and he was of infinite use in fortifying the conclusions I derived from my comparative

Sections; by his admirable and ready knowledge of the characteristic fossil species we obtained from them. The hypothesis (*supra*, page xx), either of an overlap somewhere along the unconformable junctions of the Cambrian and Silurian rocks of Denbighshire, or of a mistake committed by myself in naming some of the fossiliferous groups which appear near that junction, was proved at once to be without meaning. I had made no mistakes of the kind, nor did we find any great mistake in any of my old sections among the mountains of Caernarvonshire and Merionethshire. My work of 1831—32 was right in principle, and withstood our renewed test. We examined in great detail the two lines of the Bala Limestone, caused by synclinal flexure, securing our work by tracing both beds along their strike, and in this way we demonstrated, that the more eastern limestone bands in the Llanwddyn valley were identical with the eastern bands that cross the road, between Bala and Llangynog, as before stated. We also carefully mapped a part of the country east and north of the Northern Berwyns; and we completed in great detail sections which connected the Silurian rocks south of the Tannat, and north of the Ceiriog, shewing the emergence of the old Cambrian rocks which pass through the intervening country and form the highest crests of the Berwyns. We also examined the great fault S.E. of Llanwddyn, which produces an entire inversion of the strata through a range of several miles. This fact I had first observed in 1832, and had verified it by following the inverted beds along their strike till they had regained their normal position, and we found that we had no corrections to make in this portion of my old Sections of 1832. I mention these facts only to shew how conscientiously our work was done. We sought the truth, and would have embraced it, to whatever conclusions it might lead us.

The work done by Mr Salter and myself in 1842-43 seemed to bring to a happy end my labours among the higher mountains of North Wales; for I had re-examined all the essential parts of my old Sections, and all my groups of strata, assisted by Mr Salter in the field, and still more in the closet, by his lists of the fossils we had collected. There was no great or fatal mistake in any of my older details, and we came away rejoicing in the thought that we had done our work effectually and to a good purpose. But a very hard task remained: how were we to join our detailed work to that of the Silurian System? It appeared evident, at a glance, that the two were on some points incompatible. If our work were true, there must be some very great error lurking among the Lower Silurian Groups. How was it to be discovered? I meant to have undertaken the task myself the next year: but a serious illness compelled me to spend the summer of 1844 at one of the German baths, and in no part of that summer, or of the autumnal months, was I capable of taking the field as a Geologist. But my young friend and fellow-labourer, Mr Salter, had, with Mr Sowerby's consent, a commission from myself to examine the Llandeilo Flags

North of Builth, and to ascertain their relation to the chain of Mynedd-Epynt; and especially to ascertain whether the Llandeilo Flags were to be placed above or below the conglomerates of Dol-Fan, which ran into a remarkable chain, now I believe regarded as upper Llandovery rocks. He was then a very youthful observer, and had not learned to trust himself, when the phenomena before him seemed to contradict the opinions of those whom he considered of high authority. He brought back to me, however, at the end of the summer, a very elaborate Report, from which after its perusal I could derive no definite result; and some years afterwards it was returned to its author, who confessed that it was erroneous, and I believe destroyed it.

I spent the whole summer following (1845) in going over a part of my old work in Cumberland, Westmorland, and North Lancashire, endeavouring to bring the rocks above the Coniston Limestone (the equivalent of the Bala) into some accordance with the Groups of the Upper and true Silurian System. There could be no doubt that the Limestones of Bala and Coniston were of the same age. The fossils were numerous and almost identical in species. It was equally certain that the highest groups of the Westmorland Slate rocks, that overhang the Valley of the Lune near Kirkby Lonsdale, were on a parallel with the upper Ludlow rocks, as seen near the banks of the Towey: but how to bring the intervening groups into strict comparison with the successive upper Silurian Groups, was a task which I have never, to this day, performed to my entire satisfaction.

I mention these facts in their order, with no motives I trust of personal vanity; but to prove with what steady perseverance I went on with the task that was before me.

My object, from the first, had been to write a general work upon all the Palæozoic Rocks of England and Wales; and with this object still in view I went on from year to year, accumulating materials, which at length became too much for my sustaining powers. I had been much interrupted for many successive years by attacks of suppressed gout, and by very alarming attacks of congestion of the head; and at length the infirmities of old age had gathered round me before I had put my work in order. I will, however, leave this digression and come back to my Cambrian task.

So far my present Preface has been associated with many happy and bright remembrances—social and physical.

What is about to follow will be less satisfactory to the reader; and will be associated in my memory with acts which very painfully affected me with involuntary distrust of some whom I had counted among my best and dearest friends, and threw a moral shade over all the latter years of my Geological life.

It will perhaps be said that after the death of Sir Roderick Impey Murchison and Mr Warburton, it is wrong to revive the controversy I had with them, as I have already published my vindication in the Introduction to McCoy's Synopsis. But very few men indeed ever have an opportunity of reading that Introduction; for it is a portion of a large and expensive volume, which has very seldom been purchased since the wide diffusion of the volumes of the Palæontographical Society. On the contrary the statements of my opponents are to be seen in the *Proceedings* and *Quarterly Journal of the Geological Society*, which form part of the Library stock of every country town in which Geology is held in practical honour.

Moreover, the controversy has recently been revived with great spirit and with great talent by Prof. Sterry Hunt, F.R.S. &c., of the Canadian Survey; and I think that, under such circumstances, it would be a shameful act of moral cowardice not to speak out in my own vindication, when I can do so in the simplest words of truth and reason.

I will then do my best to state the historic truth in all simplicity and without favour or affection. Not to speak plain truth of those who are dead, while engaged in a personal vindication of truth involving questions of fact, would be destructive of the very essence and marrow of all history.

First then regarding Sir R. I. Murchison. He attended several, and I think all, of the Meetings during 1842—43, when the papers by Mr Salter and myself were read before the Geological Society: but, so far as I remember, he never made a single remark or comment during these long readings, though the subjects discussed in them affected his own works as much as mine. But very soon after the final reading of the Papers (towards the end of the year 1843) a geological map was published in his name, in which he had brushed out of sight, under a deep Silurian colour, every trace of my previous work in North Wales. This was done so quietly and silently that I never heard one whisper of it till the fact was made known to me by Mr Knipe (I think in 1851), when he called on me with a newly coloured Geological Map of England, which he had on sale. The exact date is not however material to my present purpose. Was this right or was it wrong? and was it for the interests of truth in Science? On this subject I make no further remark, but refer the reader to the Introduction to McCoy's Synopsis, and to a Paper published by myself in the *Quarterly Journal of the Geological Society* in 1852.

About the same time that Murchison had thus completed his new colouring of the map of Wales*, Mr Warburton, then President of the Geological Society, most kindly, as I thought, offered to reduce the successive communications of Mr Salter and myself, embracing the labours of the two preceding summers, into a state fit for publication. Certainly my Papers

* This map was not, I believe, published in any work connected with the Geological Society; but in the Atlas of the Society for the Diffusion of Useful Knowledge, where it lurked I believe out of my sight for more than seven years

required revision. Each of them had been written by myself in a slovenly and hasty manner; and must at least be united and copied out again before they could be printed. Most willingly therefore all my papers, and all my sections greatly improved by the graceful touch of Salter's pen, and all his own beautiful sections and sketches, were placed without reserve in the hands of Mr Warburton. But what took place after this surrender of the papers? Mr Warburton commenced his task of Reduction and very soon became involved in difficulties (as I learned from notes of enquiry sent by himself) obviously arising out of his want of knowledge of the physical structure of North Wales; and I entreated him to send me the proof-sheets, that I might be sure he understood the drift and meaning of the papers he held in charge. But he refused me the sight of any single proof-sheet, though I applied to him again and again, with increased energy after his repeated denials.

At length the Reduction was printed in the *Proceedings of the Geological Society;* and afterwards in the first volume of the *Quarterly Journal.* The sections were so much obscured by a complicated notation, which I never well understood, and by the minuteness of the scale of their Reduction, that I was never able completely to comprehend any one of them.

All our new Sections on the east side of the Berwyns were so mutilated as to be quite worthless: and instead of reproducing any of the elaborate and accurate work we had traced upon the map of the Government Survey, he first produced the Reduction of a worthless map, which was drawn upon no scale, but had been sketched by a provincial artist to illustrate a private lecture. A second map, in illustration of my papers, which appeared soon afterwards in the first volume of the *Quarterly Journal* of the Geological Society, was practically very little better than the former; and it was so overcrowded by ill-understood details as to be almost worthless. I did however hope that my original Papers, and the Sections jointly made by Salter and myself would be, in the end, returned, agreeably to the President's promise. But it was a vain hope—The greatest number of our Papers and Sections were never returned at all; and the few pages of manuscript text which did come back to me were all in the same state of mutilation, which made them absolutely useless for any purpose of verification.

It is no easy matter to explain an overbearing treatment such as I have described: but I believe Mr Warburton undertook his task for the express purpose of bringing my Papers into harmony with Murchison's scheme of covering all the older rocks of North and South Wales with Silurian colours. For in his Reductions he again and again contrived to *change my language, and make me write in a new Silurian tongue.* Was this fair and honest dealing with me?

I do not venture to affirm that Sir R. I. Murchison was a party to this unwarrantable dealing of Mr Warburton; but he unquestionably was ready to turn it to his own profit.

They were, at the time, in the closest daily communication; and it is also true that Murchison's expansion of his Silurian colours over all the older rocks of Wales, and Warburton's strange mangling of our Papers and Sections, took place very nearly at the same time; namely, just after our Communications to the Geological Society respecting the work done in 1842—43 had been completed.

With all the faults of the Reductions it was obvious that Mr Warburton had laboured hard at my Papers and Sections; and perhaps done his best to put them into a systematic form; and on that account I was willing, after I got over my first sorrow (and it was a *very great* sorrow, to endure the loss of perhaps the best two years' labours of my Geological life), to excuse some of his blunders, and to overlook the overbearing manner in which he had treated me. The case seemed without remedy, and I made no further movement in connection with it; and the matter would probably have passed away without any further notice from myself, had I not after the lapse of about 7 years received that information from Mr Knipe to which I have alluded in a former page (*supra*, p. xxvi). By that information I was at once convinced that I should be wanting in moral courage, and fail in doing what the truth of history required of me, if I did not claim my right position, as the first interpreter of the Cambrian Sections. And with these feelings I recorded in a Paper, read before the Geological Society in the year 1852, the result of a new examination of my original Papers, and a condensed abstract of what I had written connected with a previous controversy with Sir R. I. Murchison; and upon these historical details an argument was built which appeared to me incontrovertible*.

While writing under such circumstances some little excess of temper might I think have been expected. But in my present judgment, formed after a re-examination made in the calmness and serenity of old age, there was no want of temper in my Paper. It was full of matter, and I think fairly argued. It was, however, *very ill received* by the Geological Society; and all who took a part in the proceedings of the evening seemed to make it a point of honour to maintain every position which had been claimed in the works of Murchison. A week or two after the reading of this Paper I received a formal notification from the Secretary of the Geological Society, that the Council had passed a decree to extrude my Paper absolutely from their *Quarterly Journal.* They soon, however, found that this suppression of my Paper was impossible; for the new Volume, containing the offending Paper, had found its way partially before the public, and I had received the usual Author's presentation copies of my Paper.

* Murchison had previously claimed the Bala Group as Silurian, against which I entered my protest; and after some discussion I offered a compromise: viz. that of calling the Bala Group—Cambro-Silurian. This compromise was rejected by Murchison, and was afterwards withdrawn by myself.

The Council, however, repeated their blow in another form; and they passed a Resolution whereby I was forbidden to bring before them any Paper involving the Classification and Nomenclature of our older Palæozoic rocks. I thought that a Resolution, so unwise and so deeply injurious to myself, could never be sternly acted on; and would perhaps soon be forgotten. But experience taught me the contrary. It was acted upon with stern severity; and I was, after the expiration of about two years, compelled to withdraw from the Meetings of the Society; which I could not attend with any proper regard to my own honour while such a personal stigma was allowed by its Council to remain on my name.

I ought at the moment to have struck my name from its lists; but I could not bear the thought of taking a final leave of a Society, which for many years was almost my home in London, and in which I had spent many of the happiest hours of my life, and formed some of my most cherished friendships*.

The attempt to suppress my Paper, which had been subject to all previous formalities of Reference, and had actually passed through the Press, was a personal stigma unexampled, I believe, in the history of any other Philosophical Society in London.

More than twenty years have, I believe, passed away since the bitter Resolution of the Council was recorded in the books of the Geological Society, and most of my opponents have been removed by the hand of death from the good and evil of this world. But there is still a Council of the Geological Society, far removed from the feelings of irritation (whether just or unjust) which produced the stern censure of the Paper to which I have alluded; and I venture to challenge them, though now in a feeble voice, to re-peruse my old Paper and to produce from it a single paragraph or sentence which was unfit for me to write, or for the Council to read, and in any way justified the condemnation that had been passed upon it.

It is to me a thought full of melancholy, and of misgiving for the cause of honest truth, when I find that some of our best Geologists are even now vainly contriving, by buttresses and underpinnings, to lend support to the lower sections of the Silurian System; which were untrue to nature from the beginning, both in their whole conception and in the elaboration of their details: and doing this while they turn their faces away from

* After the attacks alluded to above, I continued occasionally to attend the Meetings of the Geological Society, and read one or two papers before them on different subjects. I did however twice trespass on the forbidden ground of Palæozoic Nomenclature. My first offending paper was arrested by the President (Professor Forbes was, I think, that evening in the Chair) as touching on forbidden matter. Lastly in October, 1854, a paper, of which I was the author, was partially read before the Society; but an essential part of it was not read that evening. When this came to my knowledge I withdrew the paper; and it was afterwards published (I think *in extenso*) in the *Philosophical Magazine* and *Annals of Philosophy.* This led the President, Mr Hamilton, into a severe comment, to which I replied in the above-named Journal; and so ended for ever a personal connection with the Geological Society, which ought to have ended two years sooner.

another System which had fixed its base in truth and reason, which had perfect Geographical congruity, and undoubted priority of date; and which appeals to them still in a language they cannot misunderstand, if they will read their lesson in the very order in which the Author of Nature has recorded it.

It may perhaps be objected to me that all or nearly all that has been stated in this Preface has appeared before in the Introduction to M^c^Coy's Synopsis, or in the various Papers which have been printed in my name; especially in the *Proceedings* and *Quarterly Journal of the Geological Society*, or in the *Philosophical Magazine* and *Annals of Philosophy*. But certainly they were never printed before in a form so connected and historical as in the statements of this Preface. With the solemnity becoming my old age and a conviction, forced upon me by my infirmities, that I shall never again be able to address the Public upon any subject connected with the scientific labours of my former life, I dare to affirm that the Geological Society had not a more true-hearted and loyal member than myself. The stigma fixed upon me by the Council of the Geological Society was the greatest sorrow of my old age. I never endeavoured to deprive any brother Geologist of what was his due, nor did I claim for myself any scrap of knowledge for which I could not make good my title. My Maps and Sections literally were public property, Prof. Phillips had the use of my field-work in Cumberland, when he was preparing one of his Volumes and Maps for Publication; and Mr Greenough had my Papers and Maps in his possession for weeks together, and on several successive occasions. Such details may seem but ill fitted for the pages of this Preface; but I write as one who has endured, and is still enduring, the unmerited censure of a scientific body; and who for the last time is writing in defence of his conduct as an author.

At what time the grand mistakes in the fundamental Sections of the Silurian System were first discovered, and by whom first published, is to me still unknown. I gradually made out the mistakes for myself, after clinging to the first typical Lower Silurian Groups longer than I ought to have done. I never was in the real confidence of my old companion and fellow-labourer after we parted in 1834 at the quarry, in the Bala Limestone, at the Western base of the Northern Berwyns (*supra*, p. xix).

Conclusion.

Having finished all that can with propriety be called a Preface to the following Catalogue, I will endeavour to address a few words to the resident members of the University. I can never again hope to address the Public at the length I have done in this Preface; for I feel the infirmities of old age, yearly, I might almost say daily, pressing more and more upon me, and bringing me nearer, sensibly nearer, to my last resting-place in this world. I write not sorrowfully or despondingly. I wish to address my dear and honoured friends in Cambridge in words of hope and cheerfulness. But first of all let me thank the Author of my being for having so long upheld my life in heart and hope since I first began my residence in this University. There were three prominent hopes which possessed my heart in the earliest years of my Professorship. First, that I might be enabled to bring together a Collection worthy of the University, and illustrative of all the departments of the Science it was my duty to study and to teach. Secondly, that a Geological Museum might be built by the University, amply capable of containing its future Collections; and lastly, that I might bring together a Class of Students who would listen to my teaching, support me by their sympathy, and help me by the labour of their hands. It now makes me happy to say, that all these hopes have for many years been amply realized.

It is to me no small delight to look back on the many past years when the Heads of Colleges were my sole Auditors; when we held our annual and most cheerful festive meetings on the first of May; and rejoiced over a dinner, very sumptuously provided by the Vice-Chancellor, in accordance with the express words of Dr Woodward's Will. And we all acknowledged our Founder's judgment in this festive clause of his Will. For it greatly helped to preserve a Collection, made by him in the seventeenth century, in that integrity in which it is seen to this day in one of the closets of our Museum.

On these occasions it was my duty to expound to my Auditors the annual additions made to our Collection, and the necessity there was for a more ample Museum. I was fed by good hopes, and (like many others who have tasted that food) I had to feed only upon hopes, so far as regarded the new Museum, for more than a quarter of a century. But my labour was its own reward. It gave me health, and led me into scenes of grandeur, which taught me to feel in my heart that I was among the works of the great Creator, the Father of all worlds, material or moral; and the Ordainer of those laws out of which spring all phenomena within the ken of our senses or the apprehension of our minds. I know there are men who deny the sound teaching of this lesson; but I thank God that I had been taught, from my early life, to accept these lessons as a part of God's truth; and it was my delightful task to point out year by year to my Geological Class, the wonderful manner in which the materials of the Universe were knit together, by laws which proved to the understanding and heart of man, that a great, living, intellectual, and active Power must be the creative Head of the sublime and beautiful adjustments and harmonies of the Universe.

Still nearer to us, and on that account more impressive, are the adaptations of organ to organ in every living being, great or small: and in all their complexity still governed by law, and most nicely adapted to material nature, and to all the subtle elements within which God has placed them.

How feeble I always felt myself whenever I touched upon subjects such as these! But my Class always heard me with respectful attention: for I did not introduce these subjects too often, nor did I dwell upon them too long. And I sometimes ventured to conduct my Class to thoughts of a still higher aim, connected with that being—Man—the last in order of creation, and made in the image of the Author of his being. Seeing that Man has the gift of prescience (small it may be, and vanishing from thought at once when we think of the Omniscient Prescience of God)—that he can design and contrive implements for his own use and of the nicest skill, which will give him new powers over material nature, and make him acquainted with things furthest removed from the ken of sense—the greatest and the least things accessible to the sight of man—that he has the capacity of abstract thought, and is capable of forming language, and making others understand it—that, using this as an implement of imagination, he can evolve thoughts which act upon the most powerful emotions of the heart, and fill the soul with images of glory—that he can invent another language of a mighty but far different power, which shuts out the imagination, and deals only with the abstractions of pure reason—and that through the might of this new language, and working with it among the elements of pure reason, he can logically grasp results inaccessible to any other implements of human thought—that he can tell the ever-enduring speed at which light (the first-born of heaven) travels through astral space, and count the number of its waves—that working with this logic of pure reason he can tell the astronomer to lift up his telescope to a certain point in the sky and there behold a planet never before seen by the eye of man.

That Man in his animal nature is to be counted but as one in the great kingdom of things endowed with life, we at once admit; but that in the functions and powers of his intellect (here just touched on by my feeble hand) he is absolutely removed from any co-ordination with the lower beings of Nature, is, I firmly believe, one of the most certain of well apprehended truths. We all admit that Nature is governed by law: but can we believe that a being like man is nothing but the final evolution of organic types worked out by the mere action of material causes? How are such organic evolutions to account for our sense of right and wrong, of justice, of law, of cause and effect, and of a thousand other abstractions which separate man from all the other parts of the animal world; and make him, within the limits of his duty, prescient and responsible.

The facts and sentiments connected with that which marks Humanity,—the works of man's hands, the visions of his eyes, the aspirations of his heart—appear to me utterly abhorrent from the dogmas of materialistic Pantheism. I never could be content, while thinking of such things, to feel myself dangling in mid-air without a resting-point for the sole of my foot. The true resting-point is a reception both in heart and head of a great First Cause—the one God—the Creator of all worlds, and of all things possessing life. Here we have found a true resting-place and heart's content; and so we are led to feel the sanctity and nobility of Truth, under all the forms in which it shews itself, to rejoice in its possession, and to honour it as the gift of God.

What does the Pantheist give us? A day of uncertain light, of uncertain joy, and a night of eternal darkness. But a better teaching tells us that there is a God who is the Father of the universe, and careth for all His creatures: and if we have listened to a still higher teaching, we can believe that as all the world of Nature has been progressive, so the life of man, and the labours of man, are not to end here, but are to lead him to a brighter and more glorious existence. And there is a higher teaching still, very near to us, even in our own heart and conscience:—an emanation of holy light from the Fountain-head of all light—toward which I am permitted but to take one glance while winding up this concluding address. And may our Maker grant that His holy light may guide the steps and warm the hearts of all who read this Preface!

A. SEDGWICK,

THE PRECINCTS, NORWICH.

September 17th, 1872.

The first portion of this Preface amounting to about 12 pages was dictated at Cambridge to my young assistant Walter Keeping. The remaining part, excepting the Conclusion, was written at Norwich by my servant from my dictation. The Conclusion I dictated to my Niece. Without such kind help I could have done nothing. The University will, I hope, pardon the long delay (very painful it has been to myself) in the publication of the following Catalogue.

The Comparative Schemes of Classification so kindly and clearly given by Professor Morris, and alluded to above (page xv, l. 2), ought to have been printed *in extenso* from the first; but they will appear among the papers so generously added by him to this Volume.

A TABLE

SHOWING

THE CLASSIFICATION OF THE LOWER PALÆOZOIC ROCKS.

CLASSIFICATION OF THE

Division	Sub-division	SEDGWICK.	Division	MURCHISON, 1868.	JUKES, 1863.
		Ledbury Shales.		Passage Beds.	
SILURIAN.		Downton Sandstone, Bone Bed, and	UPPER SILURIAN.	Tilestones, and	UPPER SILURIAN.
		Upper Ludlow.		Upper Ludlow.	
		Aymestry Limestone, and		Aymestry Limestone.	
		Lower Ludlow.		Lower Ludlow.	
		Wenlock Group.		Wenlock Limestone and Shale.	
		Lower Wenlock Group.		Woolhope Limestone and Shale.	
		May Hill Group.		Upper Llandovery.	
CAMBRIAN.	UPPER.	Upper Bala Group.	LOWER SILURIAN.	Lower Llandovery.	LOWER, OR CAMBRO-SILURIAN.
		Middle Bala Group.		Caradoc and Bala Rocks.	
		Lower Bala Group.		Upper and Lower Llandeilo Rocks.	
		Arenig or Skiddaw Group.			
	MIDDLE.	Tremadoc Group.			
		Ffestiniog Group.		Lingula Flags, or	
		Menevian Group.		PRIMORDIAL SILURIAN.	
	LOWER.	Harlech Group.			CAMBRIAN.
		Longmynd, Bangor, and		CAMBRIAN.	
		Llanberis Group.			

In a Table of Strata prepared by Mr H. W. Bristow (1872), the Upper and Lower Llandovery beds, considered by Sir R. I. Murchison as forming an intermediate or Middle Silurian, are kept distinct. The Lower Llandovery, Caradoc, Llandeilo, Tremadoc, Lingula and Menevian beds forming the Lower Silurian. The Graptolite shales being equal to the Arenig or Stiperstones. In the Lake district the Kirby Moor flags are equivalent to the Tilestones and Upper Ludlow, and the Bannisdale beds represent the Aymestry limestone, Lower Ludlow, Wenlock and Woolhope series. The Coniston Grits and Flags are equal to the Denbighshire Flags and Grits. The Stockdale slates (Graptolite mudstones) being equal to the Tarannon shales. The Coniston limestone is equivalent to the Caradoc or Bala beds. The Green slates and Porphyry are equivalent to the Upper Llandeilo, and the Skiddaw slates to the Arenig beds or Lower Llandeilo.

LOWER PALÆOZOIC ROCKS.

LYELL, 1871.		PHILLIPS, 1855.
UPPER SILURIAN.	Downton Sandstone and Bone Bed.	
	Upper Ludlow.	UPPER SILURIAN.
	Lower Ludlow.	
	Wenlock Limestone and Shale.	
	Woolhope Limestone and Grit.	
	Tarannon Shales.	
	Upper Llandovery, or May Hill.	
	Lower Llandovery.	LOWER SILURIAN.
LOWER SILURIAN.	Bala and Caradoc Beds.	
	Llandeilo Flags.	
	Arenig, or Stiper-Stones Group.	UPPER CAMBRIAN.
UPPER CAMBRIAN.	Tremadoc Slates.	
	Lingula Flags.	
LOWER CAMBRIAN.	Menevian Beds.	LOWER CAMBRIAN.
	Longmynd Group, including	
	Harlech Grits and Llanberis Slates.	

GEOLOGICAL SURVEY, 1865.	
UPPER SILURIAN.	Tilestones.
	Upper Ludlow.
	Aymestry Limestone.
	Lower Ludlow Beds.
	Wenlock Limestone.
	Wenlock Shale with Sandstone.
	Woolhope Limestone and Shale.
	Denbighshire Grits.
	Tarannon, or Pale Shales.
	Upper Llandovery Rock, or May Hill Sandstone.
LOWER SILURIAN.	Lower Llandovery Rock, &c.
	Caradoc or Bala Sandstone, and Limestone.
	Llandeilo Flags and Limestones.
	Graptolite Shales and Slates.
	Tremadoc Slates.
	Lingula Beds.
CAMBRIAN.	Harlech, Llanberis, St David's, and Longmynd Grits, and Conglomerates with Pale and Green Slates.

In a communication to the Geologists' Association (June 1872), on the Classification of the Cambrian and Silurian rocks, Dr Hicks nearly followed that adopted by Sir C. Lyell.

The *Lower Cambrian*, to include the Longmynd (Harlech grits and Llanberis slates and the rocks at Bray Head) and the Menevian groups.

The *Upper Cambrian*, to include the Lingula flags (Lower, Middle and Upper, called also the Maentwrog, Ffestiniog and Dolgelly) and the Tremadoc groups.

The *Lower Silurian*, to comprise the Lower and Upper Arenig, the former being a connecting link between the Tremadoc and the true Arenig rocks, the Upper and Lower Llandeilo and the Bala or Caradoc groups.

The *Upper Silurian*, to consist of the Lower and Upper Llandovery, the Wenlock and Ludlow groups.

SUMMARY OF THE CONTENTS OF THE CATALOGUE,

BY PROFESSOR MORRIS.

Cambrian.

The Lower Cambrian, including the Longmynd and Harlech groups (Sedgwick), is represented by but few forms in the collection. These are the Oldhamia, only hitherto found in Ireland, some Annelida from the Longmynd, a few Brachiopoda (Lingulella, Obolella), and a few Trilobites, which occur low down in the Longmynd group at St David's Promontory, South Wales, as Conocoryphe, Paradoxides, Microdiscus, and the interesting genus Plutonia, having affinities with Paradoxides and Anopolenus, which appears to be restricted to this zone, and, next to Paradoxides Davidis is the largest Trilobite found in the British Cambrian rocks.

The Middle Cambrian comprises the Menevian, Ffestiniog, and Tremadoc groups.

The fauna of the Menevian, or Lower Lingula flags, is represented by Trilobites of the genera Conocoryphe, Agnostus, Olenus, Paradoxides, Microdiscus, Erinnys, Anopolenus, Holocephalina, the last three genera being at present characteristic of this zone; Paradoxides and Microdiscus here become extinct, while the first three genera range into the zones above: with these are found a few Phyllopoda (Primitia, Hymenocaris), some Brachiopoda and Pteropoda, (Theca, Stenotheca, Cyrtotheca). The Pteropoda appear to be tolerably abundant in these primordial rocks, in which occur also a Cystidean (Protocystites), and some sponges.

The Ffestiniog group, or Middle and Upper Lingula flags, is chiefly represented by some Annelids and Trilobites. The genus Olenus here attains its maximum numerical development, and a species of the allied genus Dikellocephalus also occurs, together with a few Brachiopoda belonging to the genera Lingulella, Obolella and Orthis.

The fossil forms of the Lower and Upper Tremadoc groups chiefly comprise Phyllopoda and Trilobites: among the latter are the genera, now first noticed, Niobe (intermediate to the genera Asaphus and Ogygia), Psilocephalus, a very abundant form, allied to Illænus, Angelina (allied to Olenus), the most abundant of the Tremadoc Trilobites, and species of Asaphus, Ogygia, and Cheirurus. There are also some Pteropoda and Heteropoda, as Theca,

Conularia, and Bellerophon, and a species of Orthoceras, which at present is the oldest known form of the Cephalopodous group of Mollusca.

The Arenig or Skiddaw group, which is classed as the upper part of the Middle Cambrian, or as forming the base of the Upper Cambrian, contains many species of Graptolitidæ, belonging to the genera Graptolites, Diplograpsus, Phyllograptus, Didymograpsus, Dichograpsus, Tetragrapsus, and Dendrograpsus, a few worms, some genera of Trilobites of which, species of Calymene, Æglina, and Ogygia, are the most abundant, together with a few Brachiopoda, Lamellibranchiata and Pteropoda.

The Upper Cambrian comprising the Lower, Middle and Upper Bala groups, is represented in the collection by a numerous fauna. There are many species of Graptolites, Corals (both Tabulate and Rugose), Brachiopoda, and Trilobites, which latter here attain their maximum developement; together with a less number of species of Lamellibranchiata, Gasteropoda, Pteropoda and Cephalopoda, and some Crinoids and Starfishes (Protaster, Palæaster).

Silurian.

The May Hill Sandstone or upper Llandovery is represented by some species of Corals belonging to the genera Favosites, Heliolites, and Petraia; a few Annelids and Trilobites: the Brachiopoda, however, are the most abundant, whilst the species of Gasteropoda, Lamellibranchiata and Cephalopoda are but few in number.

The Lower Wenlock group. In this, as in the preceding group, the Graptolites have diminished in number, and the Corals, Echinoderms and Trilobites have but few representatives; but, as in the preceding period, the Brachiopoda are most abundant, the Cephalopoda also comprising a large number of species.

The Wenlock group contains a rich fauna: the Bryozoa and Actinozoa, both Tabulate and Rugose, are here numerically abundant. The Crinoids and Cystideans contain many species, as also do the Trilobites (40), Brachiopoda (82), Lamellibranchiata (31), Gasteropoda (42), Cephalopoda (32), together with some few Heteropoda and Pteropoda.

Lower Ludlow and Aymestry Limestone. This group contains two or three species of Graptolites, a few Corals, Crinoids, Phyllopoda, and Trilobites, some species of Starfish (Asteroidea) belonging to five genera, and some Crustacea, belonging to the order Merostomata (Slimonia, Pterygotus), which, although represented in the preceding Llandovery and Wenlock groups, increases in numbers in the lower and upper Ludlow rocks. The species of Brachiopoda and Lamellibranchiata are far less in number than in the preceding Wenlock group, and are associated with some forms of Gasteropoda, Heteropoda, and Cephalopoda, belonging chiefly to the genera Orthoceras, Phragmoceras and Trochoceras.

The Upper Ludlow mudstones contain representatives of most of the classes previously mentioned, the Merostomata affording species of Eurypterus and Pterygotus, as well as the interesting genus Hemiaspis; the Lamellibranchiata, among the Mollusca, comprise the largest number of specific forms, while the Cephalopoda contain about the same number of species (16), as in the preceding group.

The Ludlow bone-bed and Downton sandstone are represented by few forms, and these chiefly belong to the Molluscan classes, for Trilobites have become rare, and Corals and Echinoderms are absent, there are however, a few Merostomata (Pterygotus, Hemiaspis), and Phyllopoda (Beyrichia, Leperditia). The bone-bed is the marked feature of this group, with its numerous fragmentary remains of fish. The Ledbury shales contain but few species, which chiefly belong to the Merostomata and Phyllopoda.

The Catalogue enumerates about 910 named species, but many other forms are noticed, together with their localities, to which specific names are not assigned.

TABLE

SHEWING THE RANGE OF THE GENERA AS INDICATED BY THIS CATALOGUE.

	Cambrian.								Silurian.						
	Lower Cambrian.	Menevian.	Ffestiniog.	Tremadoc.	Arenig Group.	Lower Bala.	Middle Bala.	Upper Bala.	May Hill Group.	Lower Wenlock.	Wenlock Group.	Lower Ludlow.	Upper Ludlow.	Downton Sand-stone.	Ledbury Shales.
Plantæ															
Oldhamia	*														
Chondrites	...	...	...	...	...	...	...	...	...	...	*				
Spongarium	...	...	...	...	...	...	...	...	...	...	...	*	*		
Pachytheca	...	...	...	...	...	...	...	...	...	...	...	...	...	*	
Actinophyllum	...	...	...	...	...	...	...	...	...	...	...	...	...	*	*
Amorphozoa															
Protospongia	...	*													
Astylospongia	...	...	...	...	...	...	*								
Stromatopora	...	...	...	...	...	...	*	...	*	...	*				
Sphærospongia	...	...	...	...	...	...	*	...							
Ischadites	...	...	...	...	...	...	*	...	...	...	*				
Vioa	...	...	...	...	...	...	*	...	*						
? Nidulites	...	...	...	...	...	...	...	*							
Cnemidium	...	...	...	...	...	...	...	...	...	...	*				
Verticillopora	...	...	...	...	...	...	...	...	...	...	*				
Pasceolus	...	...	...	...	...	...	...	...	...	...	...	...	*		
Tetragonis	...	...	...	...	...	...	...	...	...	...	...	...	*		
Hydrozoa															
Graptolithus	...	...	...	...	*	*	*	...	*	*	*	*	*		
Rastrites	...	...	...	...	...	*									
Diplograpsus	...	...	...	...	*	*	*								
Phyllograptus	...	...	...	...	*										
Didymograpsus	...	...	...	...	*	*									
Tetragrapsus	...	...	...	...	*										
Dichograpsus	...	...	...	...	*										
Dendograpsus	...	...	...	...	*	...	...	...	...	...	...	*			
Climacograpsus	...	...	...	...	...	*									
Protovirgularia	...	...	...	...	...	*									
Retiolites	...	...	...	...	*	...	...	...	...	*					

f

	Cambrian.								Silurian.						
	Lower Cambrian.	Menevian.	Ffestiniog.	Tremadoc.	Arenig Group.	Lower Bala.	Middle Bala.	Upper Bala.	May Hill Group.	Lower Wenlock.	Wenlock Group.	Lower Ludlow.	Upper Ludlow.	Downton Sand-stone.	Ledbury Shales.
Actinozoa															
Stenopora	...	...	...	...	...	*	*	*	...	*	*	*	*		
Nebulipora	...	...	...	...	...	*	*	*	...	...	*	...	*		
Heliolites	...	...	...	...	...	*	*	*	*	...	*				
Favosites	...	...	...	...	...	...	*	*	*	*	*				
Halysites	...	...	...	...	...	...	*	*	*	...	*				
Omphyma	...	...	...	...	...	...	*	*	*	...	*				
Cyathophyllum	...	...	...	...	...	...	*	...	...	*	*				
Sarcinula	...	...	...	...	...	...	*	...	...	...	*				
Petraia	...	...	...	...	...	...	*	*	*	*					
Palæocyclus	...	...	...	...	...	...	...	...	*	...	*				
Fistulipora	...	...	...	...	...	...	...	...	...	...	*				
Cœnites	...	...	...	...	...	...	...	...	...	...	*				
Alveolites	...	...	...	...	...	...	...	...	...	...	*				
Chætetes	...	...	...	...	...	...	...	...	...	...	*				
Labechia	...	...	...	...	...	...	...	...	...	...	*				
Thecia	...	...	...	...	...	...	...	...	...	...	*				
Syringopora	...	...	...	...	...	...	...	...	...	...	*				
Aulacophyllum	...	...	...	...	...	...	...	...	...	...	*				
Ptychophyllum	...	...	...	...	...	...	...	...	...	...	*				
Goniophyllum	...	...	...	...	...	...	...	...	...	...	*				
Arachnophyllum	...	...	...	...	...	...	...	...	...	...	*				
Cystiphyllum	...	...	...	...	...	...	...	...	...	...	*				
Clisiophyllum	...	...	...	...	...	...	...	...	...	...	*				
Cyathaxonia	...	...	...	...	...	...	...	...	...	...	*				
Echinodermata															
Protocystites	...	*													
Echinosphærites	...	...	...	...	...	...	*								
Sphæronites	...	...	...	...	...	...	*								
Apiocystites	...	...	...	...	...	...	...	...	...	...	*				
Prunocystites	...	...	...	...	...	...	...	...	...	...	*				
Pseudocrinites...	...	...	...	...	...	...	...	...	...	...	*				
Echino-encrinites	...	...	...	...	...	...	...	...	...	...	*				
Ateleocystites	...	...	...	...	...	...	...	...	...	...	*				
Pisocrinus	...	...	...	...	...	...	...	...	...	...	*				
Glyptocrinus	...	...	...	...	...	...	...	...	...	...	*				
Periechocrinus...	...	...	...	...	...	...	...	...	*	*	*				
Actinocrinus	...	...	...	...	...	...	...	...	*	...	*				
Hypanthocrinus	...	...	...	...	...	...	...	...	...	*	...	*			
Herpetocrinus	...	...	...	...	...	...	...	...	...	*	*				
Cheirocrinus	...	...	...	...	...	...	...	...	...	...	*				
Marsupiocrinus	...	...	...	...	...	...	...	...	...	...	*				
Dimerocrinus	...	...	...	...	...	...	...	...	...	...	*				
Mariacrinus	...	...	...	...	...	...	...	...	...	...	*				
Platycrinus	...	...	...	...	...	...	...	...	...	...	*				
Crotalocrinus	...	...	...	...	...	...	...	...	...	...	*	*			
Cyathocrinus	...	...	...	...	...	...	...	...	...	...	*				
Taxocrinus	...	...	...	...	...	...	...	...	...	...	*	*			
Ichthyocrinus	...	...	...	...	...	...	...	...	...	...	*	*			

	CAMBRIAN.								SILURIAN.						
	Lower Cambrian.	Menevian.	Ffestiniog.	Tremadoc.	Arenig Group.	Lower Bala.	Middle Bala.	Upper Bala.	May Hill Group.	Lower Wenlock.	Wenlock Group.	Lower Ludlow.	Upper Ludlow.	Downton Sand-stone.	Ledbury Shales.
Echinodermata (*continued*)															
Protaster	...	...	...	...	...	...	*	...	...	...	...	*			
Palæaster	...	...	...	...	...	...	*	...	...	...	...	*			
Lepidaster	...	...	...	...	...	...	...	...	...	...	*				
Palasterina	...	...	...	...	...	...	...	...	...	...	...	*			
Palæocoma	...	...	...	...	...	...	...	...	...	...	...	*			
Bdellacoma	...	...	...	...	...	...	...	...	...	...	...	*			
Annelida															
Arenicolites	*														
Scolecites	*	...	*												
Scolecoderma	...	...	*	...	*	...	...	...	...	...	*				
Cruziana	...	...	*												
Pyritonema	...	...	...	...	...	*									
Nemertites	...	...	...	...	...	*									
Myrianites	...	...	...	...	...	*									
Nereites	...	...	...	...	...	*									
Palæochorda	...	...	...	...	*										
Tentaculites	...	...	...	...	...	...	*	*	*	...	*	...	*		
Trachyderma	...	...	...	...	...	...	*								
Serpulites	...	...	...	...	...	...	*	...	...	...	...	*	*		
Crossopodia	...	...	...	...	...	...	*	...	...	...	...	...	...	*	
Cornulites	...	...	...	...	...	...	...	...	*	*	*	...			
Spirorbis	...	...	...	...	...	...	...	...	...	*	...	*			
Turrilepas	...	...	...	...	...	...	...	...	...	...	*				
Crustacea															
Conocoryphe	*	*	*	*	...	*									
Plutonia	*														
Paradoxides	*	*													
Microdiscus	*	*													
Agnostus	*	*	*	...	*	*	*								
Holocephalina...	...	*													
Erinnys	...	*													
Anopolenus	...	*													
Arionellus	...	*													
Olenus	...	*	*												
Dikellocephalus	...	*	*												
Psilocephalus	...	...	...	*											
Asaphus	...	...	...	*	*	*	*								
Ogygia	...	...	...	*	*	*									
Angelina	...	...	...	*											
Cheirurus	...	...	...	*	...	*	*	*	*	...	*				
Trinucleus	...	...	...	...	...	*	*								
Phacops	...	...	...	...	*	...	*	*	*	*	*	*	*		
Calymene	...	...	...	...	*	*	*	*	...	...	*	*			
Dionide	...	...	...	...	*										
Trinucleus	...	...	...	...	*	...	*	*							
Ampyx	...	...	...	...	*	*									
Homalonotus	...	...	...	...	*	*	*	...	...	*	*	...	*		

	Cambrian.								Silurian.						
	Lower Cambrian.	Menevian.	Ffestiniog.	Tremadoc.	Arenig Group.	Lower Bala.	Middle Bala.	Upper Bala.	May Hill Group.	Lower Wenlock.	Wenlock Group.	Lower Ludlow.	Upper Ludlow.	Downton Sand-stone.	Ledbury Shales.
Crustacea (*continued*)															
Æglina	...	...	...	...	*	*									
Barrandia	...	...	...	...	...	*									
Lichas	...	...	...	...	...	*	*	*	...	...	*				
Illænus	...	...	...	...	...	*	*	*	*	...	*				
Acidaspis	...	...	...	...	...	...	*	...	...	*	*	*			
Staurocephalus	...	...	...	...	...	...	*	...	...	...	*				
Sphærexochus	...	...	...	...	...	...	*	...	...	...	*				
Cybele	...	...	...	...	...	...	*								
Encrinurus	...	...	...	...	...	...	*	*	...	...	*				
Harpes	...	...	...	...	...	...	*								
Proetus	...	...	...	...	...	...	...	...	*	...	*	*			
Cyphaspis	...	...	...	...	...	...	...	...	...	...	*				
Primitia	*	*	...	...	*	*	*								
Hymenocaris	...	*	*												
Ceratiocaris	...	...	...	*	...	...	*	...	...	...	*	*	*		
Lingulocaris	...	...	...	*											
Cythere	...	...	...	...	...	*									
Beyrichia	...	...	...	...	...	*	*	*	...	...	*	...	*	*	
Caryocaris	...	...	...	...	*										
Peltocaris	...	...	...	...	...	...	...	...	...	*					
Dictyocaris	...	...	...	...	...	...	...	...	...	...	...	...	*		
Leperditia	...	...	...	...	...	...	...	...	...	...	...	...	...	*	*
Pterygotus	...	...	...	...	...	...	...	...	...	...	...	*	*	*	*
Slimonia	...	...	...	...	...	...	...	...	...	...	...	*	*		
Hemiaspis	...	...	...	...	...	...	...	...	...	...	...	...	*	*	
Eurypterus	...	...	...	...	...	...	...	...	...	...	...	...	*	...	*
Bryozoa															
Glauconome	...	...	...	...	...	...	...	*							
Ptilodictya	...	...	...	...	...	...	...	*							
Trematopora	...	...	...	...	...	...	...	...	...	...	*				
Discopora	...	...	...	...	...	...	...	...	...	...	*				
Ceriopora	...	...	...	...	...	...	...	...	...	...	*				
Fenestella	...	...	...	...	...	...	...	...	...	...	*				
Dictyonema	...	...	*	...	...	...	...	...	...	...	*				
Polypora	...	...	...	...	...	...	...	...	...	...	*				
Brachiopoda															
Lingulella	*	*	*	*											
Obolella	...	*	*	...	*										
Discina	...	*	...	...	...	*	...	...	...	*	*	*	*		
Orthis	...	*	*	...	*	...	*	*	*	*	*	*	*		
Kutorgina	...	...	*												
Lingula	...	...	...	...	*	*	*	*	*	...	*	*	*	*	*
Crania	...	...	...	...	...	*	*	...	...	...	*	...	*		
Siphonotreta	...	...	...	...	...	*	...	...	...	...	*				
Strophomena	...	...	...	...	...	*	*	*	*	*	*	*	*		
Leptæna	...	...	...	...	...	*	*	*	*	*	*				
Trematis	...	...	...	...	...	...	*	...	...	...	*	...	*		

	Cambrian.								Silurian.						
	Lower Cambrian.	Menevian.	Ffestiniog.	Tremadoc.	Arenig Group.	Lower Bala.	Middle Bala.	Upper Bala.	May Hill Group.	Lower Wenlock.	Wenlock Group.	Lower Ludlow.	Upper Ludlow.	Downton Sand-stone.	Ledbury Shales.
Brachiopoda (*continued*)															
Atrypa	...	...	...	...	...	...	*	*	*	*	*	*			
Triplesia	...	...	...	...	...	...	*								
Rhynchonella	...	...	...	...	...	...	*	*	*	*	*	*	*		
Pentamerus	...	...	...	...	...	...	*	*	*	*	*	*			
Porambonites	...	...	...	...	...	...	*								
Meristella	...	...	...	...	...	...	...	*	*	*	*	*			
Obolus	...	...	...	...	...	...	...	...	*	...	*				
Spirifer	...	...	...	...	...	...	...	...	*	*	*	...	*		
Chonetes	...	...	...	...	...	...	...	...	*	*	*	*	*	*	
Athyris	...	...	...	...	...	...	...	...	...	*	*	*			
Retzia	...	...	...	...	...	...	...	...	...	*	*				
Eichwaldia	...	...	...	...	...	...	...	...	...	...	*				
Nucleospira	...	...	...	...	...	...	...	...	...	...	*				
Lamellibranchiata															
Palæarca	...	...	...	...	*	...	...	...	...	*	...	...	...	*	
Ctenodonta	...	...	...	...	*	...	*	*	*	*	*	*	*	*	
Pterinea	...	...	...	...	...	...	*	...	*	*	*	*	*	*	
Modiolopsis	...	...	...	...	...	...	*	...	...	...	*	*	*	*	
Lyrodesma	...	...	...	...	...	...	*	...	*						
Cleidophorus	...	...	...	...	...	...	*	...	...	*	*				
Orthonotus	...	...	...	...	...	...	*	*	*	*	*	*	*	*	
Cuculella	...	...	...	...	...	...	*	...	...	...	*	*	*	*	
Mytilus	...	...	...	...	...	...	...	...	*	*	...	...	*		
Nuculites	...	...	...	...	...	...	...	...	...	*					
Cardiola	...	...	...	...	...	...	...	...	...	*	*	*			
Avicula	...	...	...	...	...	...	...	...	...	...	*	...	*		
Ambonychia	...	...	...	...	...	...	...	...	...	...	*	*			
Pseudaxinus	...	...	...	...	...	...	...	...	...	...	*	*	*		
Pleurorhynchus	...	...	...	...	...	...	...	...	...	...	*				
Goniophora	...	...	...	...	...	...	...	...	...	...	*	...	*	*	
Grammysia	...	...	...	...	...	...	...	...	...	...	*	...	*	*	
Lunulacardium	...	...	...	...	...	...	...	...	...	...	*				
Anodontopsis	...	...	...	...	...	...	...	...	...	...	...	*	*		
Gasteropoda															
Murchisonia	...	...	...	...	...	*	*	*	*	*	*	*	*	*	
Pleurotomaria	...	...	...	...	...	*	...	*	...	*	*	*	*		
Raphistoma	...	...	...	...	...	*	...	*							
Euomphalus	...	...	...	...	...	*	...	*	*	...	*				
Cyclonema	...	...	...	...	...	...	*	...	*	...	*	*	*		
Holopea	...	...	...	...	...	...	*	...	*						
Holopella	...	...	...	...	...	...	*	*	*	*	*	*	*	*	
Loxonema	...	...	...	...	...	...	*	...	...	...	*	*			
Trochus	...	...	...	...	...	...	...	*	...	*					
Acroculia	...	...	...	...	...	...	...	...	*	...	*	*			
Trochonema	...	...	...	...	...	...	...	...	...	...	*				
Macrocheilus	...	...	...	...	...	...	...	...	...	...	*				
Naticopsis	...	...	...	...	...	...	...	...	...	...	...	*	*		
Platyschisma	...	...	...	...	...	...	...	...	...	...	...	...	*	*	

	Cambrian.								Silurian.						
	Lower Cambrian.	Menevian.	Ffestiniog.	Tremadoc.	Arenig Group.	Lower Bala.	Middle Bala.	Upper Bala.	May Hill Group.	Lower Wenlock.	Wenlock Group.	Lower Ludlow.	Upper Ludlow.	Downton Sand-stone.	Ledbury Shales.
Heteropoda															
Bellerophon	...	...	...	*	...	...	*	*	*	*	*	...	...	*	
Maclurea	...	...	...	...	...	*									
Pteropoda															
Theca	*	*	...	*	*	...	*	...	...	...	*	...	*		
Stenotheca	...	*													
Cyrtotheca	...	*													
Conularia	...	...	...	...	*	...	*	...	...	...	*	*	*		
Eccyliomphalus	...	...	...	...	...	*	...	*	*	...	...	...	*		
Graptotheca	...	...	...	...	...	...	...	...	...	...	...	*			
Cephalopoda															
Orthoceras	...	...	...	*	*	*	*	*	*	*	*	*	*	*	
Cyrtoceras	...	...	...	...	...	*	...	*	...	...	*	*			
Ormoceras	...	...	...	...	...	...	*								
Lituites	...	...	...	...	...	...	*	...	...	...	*	*			
Phragmoceras	...	...	...	...	...	...	...	...	*	...	*	*			
Trochoceras	...	...	...	...	...	...	...	...	...	*	*	*			
Tretoceras	...	...	...	...	...	...	...	...	...	...	...	...	...	*	
Pisces															
Onchus	...	...	...	...	...	...	...	...	...	...	...	...	...	*	*
Pteraspis	...	...	...	...	...	...	...	...	...	...	...	...	...	*	

ADDENDA.

Page 3. Protospongia diffusa, Rep. Brit. Assoc., 1865, p. 285.
Protocystites, Rep. Brit. Assoc., 1865, p. 285.
4. Agnostus Davidis, ib. p. 285.
4. Agnostus scutalis, ib. p. 285.
6. Anopolenus Henrici and Salteri, p. 481 not 478.
7. Olenus cataractes, Dec. XI., Pl. 8, fig. 14.
10. Hymenocaris vermicauda, Pl. 1, Pl. 2, figs. 1—4.
10. Agnostus princeps, line 3 from bottom (dele 6) add Pl. 4, f. 2, 11.
23. Asaphus Solvensis and A. Menapiæ belong to Niobe, according to Dr Hicks.

128. Ateleocystites Fletcheri.

In the *Geological Magazine*, 1871, Vol. VIII., p. 71, Mr Woodward calls attention to the genus Ateleocystites of Billings figured and described by him in Decade III. of the publications of the Geological Survey of Canada, Montreal 1858. In a letter which accompanies this notice, Mr Billings considers *Ateleocystites*, as suggested by Mr Woodward, to be generically identical with *Placocystites*, of De Koninck, *Geol. Mag.*, 1870[1], Vol. VII., p. 260. Pl. VII., figs. 2, 3, 4, and with *Anomalocystites*, of Hall, Pal., New York, 1859, Vol. III., p. 132, pl. 7A. and 88.

The *Ateleocystites Fletcheri* appears to be the same species as the *A.* (*Placocystites*) *Forbesianus* of De Koninck; both were obtained from the Wenlock limestone of Dudley. Figures of more perfect specimens

Ateleocystites Fletcheri, Salter = *Placocystites Forbesianus*, De Koninck.

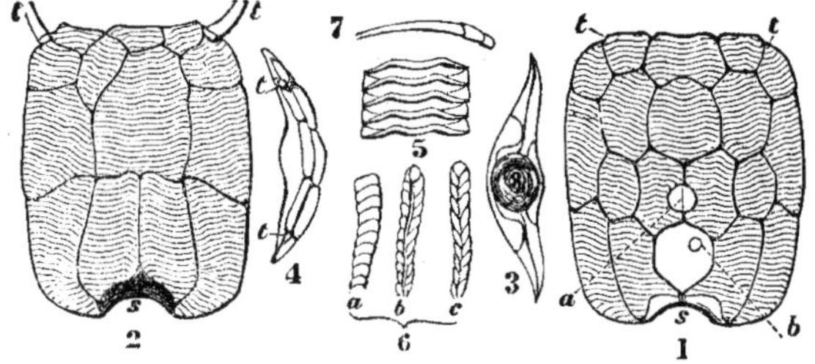

FIG. 1. Convex side, showing the so-called "anal plate" (*a*), and the ovarian pore (*b*), the base of the tentacles (*t*, *t*), and the point of attachment for the stem (*s*).
„ 2. Concave side, showing the tentacles (*t*, *t*).
„ 3. View of the lower extremity of the body, shewing the attachment of the stem (*s*).
„ 4. View of the top of the body, showing the points of attachment for the arms or tentacles (*t*, *t*).
„ 5. Portion of stem near the body: drawn from a specimen having a portion of the stem still remaining attached.
„ 6. *a*, *b*, *c*. Three views of a small tapering stem, found detached, but having the same characteristic sculpture visible upon its joints observed in *Ateleocystites*. Probably the lower extremity of the stem.
„ 7. One of the arms, or tentacles, drawn from a specimen, having the arm still attached to the body.
(All the above specimens are in the British Museum, and were obtained from the Wenlock Limestone, Wren's Nest, Dudley).

[1] Translated from the Bulletins de l'Acad. Roy. Belg., 2me Série, tome XXVIII., pp. 547—551. Planche, figs. 2 and 3. 1869.

are here reproduced by the kindness of the Editor of the *Geological Magazine.* The *Ateleocystites Huxleyi,* Billings (Canadian Fossils, Dec. III. p. 73), with which the above species has been compared, is smaller and less broad, and was obtained from the Trenton limestone. The *A. cornutus,* Hall, l. c., p. 133, Plate 7 A, figs. 5—7 is from the Pentamerus limestone of the Lower Helderberg group, and the *A. disparilis,* Hall, l. c. p. 145, Pl. 88, figs. 1—4 is from the Oriskany sandstone. Another form referred to this genus from the Upper Caradoc of Shole's Hook, Pembrokeshire, is in the collection of the Geological Survey.

Tremadoc Group, p. 15:—In a paper read before the British Association (Aug., 1872), Dr Hicks fully confirmed the occurrence of the Tremadoc rocks previously recognised in 1866, by Mr Salter and himself at Ramsey Island, St David's, where they overlie the Lingula flags, and are from 800 to 1000 feet in thickness, with numerous fossils, nearly all the species as well as many of the genera being new. They comprise Brachiopoda of the genera *Lingula, Orthis, Obolella,* and the Lamellibranchiate genus *Ctenodonta,* also species of *Orthoceras, Theca, Bellerophon,* an Encrinite and a Starfish, besides nine species of Trilobites belonging to *Niobe, Conocoryphe, Cheirurus,* and a genus allied to *Dikellocephalus.* These are followed by the Arenig group having a thickness of a 1000 feet, and contain the genera *Asaphus, Ogygia, Æglina, Trinucleus, Ampyx, Calymene* and *Agnostus;* also *Conularia, Theca, Orthoceras, Bellerophon, Lingula, Orthis.* In this group Mr J. Hopkinson has recognised more than 20 species of Graptolites, belonging to the genera,—*Didymograpsus, Tetragrapsus, Phyllograptus, Ptilograpsus, Dendrograpsus, Callograptus, Retiolites* and *Loganograptus,* and also *Dictyonema,* from which association Mr Hopkinson considers the beds to be the equivalent of the Quebec group of Canada, the Skiddaw slates of Cumberland, and the Arenig rocks of Shelve.

PALÆOZOIC SYSTEM.

GENERAL SECTION OF THE PALÆOZOIC SYSTEM IN BRITAIN.

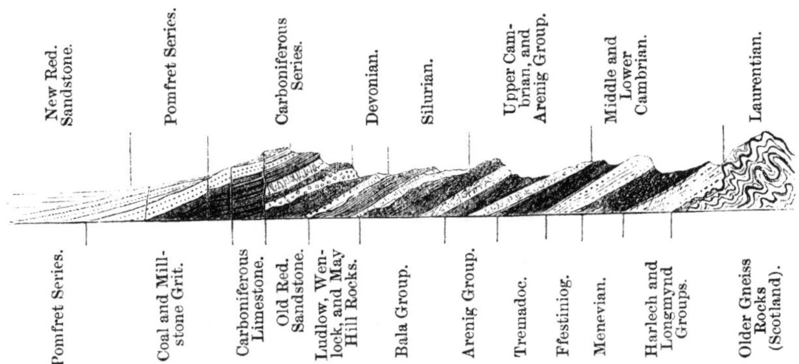

LOWER PALÆOZOIC.
1. Lower Cambrian (Sedgwick).
2. Middle Cambrian (do.).
3. Upper Cambrian (do.).
4. Silurian (Murchison).

MIDDLE PALÆOZOIC.
5. Devonian (Sedgwick and Murchison).

UPPER PALÆOZOIC.
6. Carboniferous (Smith).
7. Pomfret Series (Sedgwick. Permian, Murchison).

1. Lower Cambrian. (Sedgwick, Phillips, Lyell, &c. Cambrian of Geol. Survey).
   a. Longmynd Group (Sedgwick), certainly in great part lower than the
   b. Harlech or Bangor Group (Sedgwick).

Case and Column of Drawers.	Reference to McCoy's Synopsis: and Figures of Genera.	Names and References; Observations, &c.	Numbers and Localities.
	Marine Plants.	**CRYPTOGAMIA ALGÆ** (sea-weeds).	
G h. 1		**Oldhamia antiqua,** Forbes (Kinahan, Trans. Roy. Irish Acad. Vol. XXIII. p. 557, fig. 5). This fossil is believed by Dr. Busk and Rev. J. M. Berkeley to be a plant of the nullipore family—allied to *Acetabularia,* a Mediterranean type. Prof. Forbes thought it a Sertularian polype: but the cells are not uniform or even distinct. The radiating branchlets are connected by membrane. (See Mem. Geol. Surv. Vol. III. p. 281.)	Only known yet in two localities in Wicklow, viz. Carrickreilly Mountains, and Bray Head. a. 632*, Bray Head (presented by the Irish Survey). Longmynd Group.

* The numbers refer to tickets on the specimens. The localities and the donor's name follow the numbers.

Case and Column of Drawers.	Reference to McCoy's Synopsis: and Figures of Genera.	Names and References; Observations, &c.	Numbers and Localities.
Gh		**Oldhamia radiata,** Forbes (Kinahan, l. c. p. 557, figs. 3—5, 8—10). This has no stem, and is probably a distinct genus from the last, which has been termed *Murchisonites antiquus* by Göppert in his "Flora der Silurische und Devonische Schichten."	a. 630, a. 631, Longmynd group of Bray Head, Wicklow.—(Also from Irish Survey.)
	Worm-burrows.	**ANNELIDA** (Worm tracks and burrows).	
G Gh		**Arenicolites sparsus,** Salter (Quart. Geol. Journ. Vol. XIII. pl. 5, fig. 2). The minute exit- and entrance-holes of the burrows of marine worms on rippled sand surfaces.	a. 9, Longmynd, Shropshire. Under surface of beds. (Presented by Mr. R. Lightbody, and Mr. Salter.)
		Arenicolites didymus, Salter, ib. Vol. XII. Mouths of burrows close and parallel.	b. 257, Yearling Hill, Longmynd. (Pres. by Mr. Salter.)
	Arenicolites.	**Scolecites.** Burrows, filled up with excreta of the worm, are common in Cambrian rocks (Mem. Geol. Surv. Vol. III. 243, fig. 2).	
G (Glass case).		**Rain-imprints,** on a peculiar muddy layer, suncracked and largely rippled, have been found in more than one locality in the Longmynd (Quart. Geol. Journ. Vol. XIII. pl. 5, fig. 1).	a. 1, Yearling Hill, Longmynd. (Presented by Mr. Lightbody.)
	Trilobites. (*Low Entomostraca.* The parts of the mouth are not yet known. Name derived from the trilobed form.)	**CRUSTACEA Trilobita.** Lately discovered by Dr. Hicks, of St. David's.	
G Gh		**Conocoryphe Lyelli,** Hicks (Quart. Geol. Journ. 1869). For remarks on genus, see p. 5.	a. 471, a. 475, St. David's, 400 ft. down in the Harlech Group. (Dr. Hicks.)
G Gh		**Plutonia,** n. g. **Sedgwickii,** Hicks. Very like *Paradoxides,* but as yet undescribed. It has a narrow glabella, broad ribs (*pleuræ,* which are the first leg-joints), and a tubercular surface. Hicks (1868), Brit. Assoc. Reports.	a. 472, a. 473, a. 474, same locality. (Presented by Dr. Hicks.)
Gh		**Leperditia** (or **Primitia?**) **prima,** Hicks.	a. 283, St. David's. (Dr. Hicks.)
		Paradoxides Harknessi, Hicks MS. See p. 6.	
Gh		**Microdiscus sculptus,** Hicks MS. See p. 4.	a. 286, do. do.
Gh	*Plutonia.*	**Agnostus cambrensis,** Hicks. (Page 4 for genus.)	a. 285, do. do.
		MOLLUSCA.	
		Discina, sp. See p. 5.	These species are in Dr. Hicks' Cabinet.
		Obolella, sp. See p. 6 for the genus.	
Gh		**Lingulella ferruginea,** Salter. This and its var. *ovalis* run down at least 1200 feet into the red rocks of the Harlech Group. Quart. Geol. Journ. 1867, Vol. XXIII. p. 340, fig. 1—3.	a. 284, St. David's. (Dr. Hicks.)
	Lingulella.	**Theca,** sp. See p. 6.	

SECTION OF THE LOWER AND MIDDLE CAMBRIAN IN N. WALES.

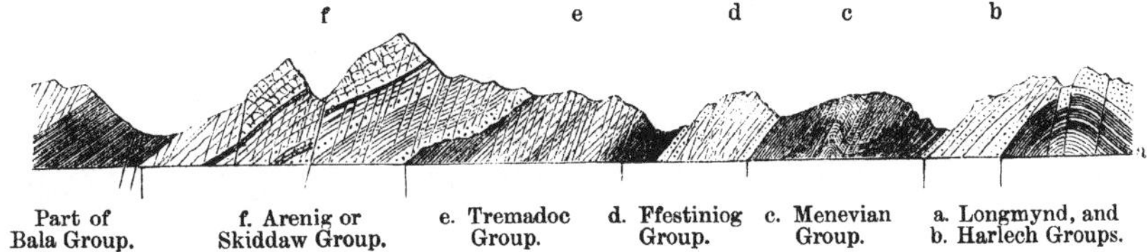

2. MIDDLE CAMBRIAN (Sedgwick). LINGULA FLAGS (Auct.). UPPER CAMBRIAN (Lyell and Salter).

c. Menevian Group (Salter and Hicks). Lower Lingula flag (Sedgwick).

d. Ffestiniog Group (Sedgwick). Middle Lingula flag and Upper Lingula flag (Salter).

e. Tremadoc Group (Sedgwick, Salter, Phillips).

c. MENEVIAN* GROUP. N. and S. Wales; Sweden; Bohemia (Etage C of M. Barrande); N. America and Canada; St. John's Group in Newfoundland.

Case and Column of Drawers.	Reference to McCoy's Synopsis: and Figures of Genera.	Names and References; Observations, &c.	Numbers and Localities.
	Sponges.	**AMORPHOZOA** (Sponges and Foraminifera).	
G h		**Protospongia fenestrata,** Salter (Quart. Geol. Journ. Vol. XX. pl. 13, fig. 12). A very simply formed sponge, with the fibres often rudimentary, and crossing at right angles on the plane surface, as minute stars.	Porth-rhaw, St. David's. (Pres. Dr. Hicks.) b. 279 (Mr. Homfray). St. David's. a. 252, Tyddyngwladis, in the Gold mines, Dolgelly. (Pres. by Mr. J. Plant, 1869.)
		[**P. flabella,** Hicks, is in Dr. Hicks' cabinet.]	
G h		**Protospongia diffusa,** Salter (Reports Brit. Assoc.). A less regular species, with closer and more matted fibres. There are probably several forms of the genus.	a. 289, St. David's. (Dr. Hicks†.)
	Cystideæ (globe-crinoids).	**ECHINODERMATA** (Crinoids, Starfish, Sea-urchins), more particularly noted in the Bala group.	
G h		**Protocystites**‡ (Salter MSS.) sp. Rather an obscure fossil, with long stem and arms. Fragments only are known, and are chiefly in Dr. Hicks' rich cabinet.	a. 281, Same locality (same donor).

* Menevia is the classic name for St. David's. The bishop is styled Episcopus Menevensis.

† The names of donors are in parentheses. Brackets [ ] include such species as are not yet obtained, but are likely to be so.

‡ This MS. name, like many others in the Catalogue, is given for present convenience, and can have no authority till the forms are truly figured and described.

Case and Column of Drawers.	Reference to McCoy's Synopsis: and Figures of Genera.	Names and References; Observations, &c.	Numbers and Localities.
		CRUSTACEA Trilobita.	
Gh	*Trilobites.*	**Agnostus princeps,** Salter (Quart. Geol. Journ. Vol. xx. pl. 13, fig. 8). A species of long range in the Lingula flag. The genus is the simplest and most rudimentary of all Trilobite forms. Only 2 segments to the body.	a. 4, Maentwrog waterfall. (Mr. Ash.) b. 273, do. (D. Homfray). a. 255, Cwm heisian, Gold mines, Dolgelly. (Mr. Plant, 1869.)
Gh		**Agnostus,** sp. Convex, lobeless, like *A. bibullatus,* Barr.	b. 266, Gwynfynydd, Dolgelly Gold mines. (Mr. D. Homfray.)
Gh	*Agnostus.* [Rudimentary trilobites, with few segments to the thorax. The characteristic "facial suture" is absent in these early forms.]	**Agnostus Davidis,** id. (Reports Brit. Assoc. 1865. Trans. Sect.) One of the largest species. Equal in size to *A. aculeatus,* Angelin, but remarkably even and slightly lobed.	St. David's. (Dr. Hicks.) a. 254, Cwm heisian; Mr. Plant. b. 271, Rhaidr ddu valley, Maentwrog. (Mr. D. Homfray.)
		Agnostus scarabæoides, Hicks. Surface wholly granular.	[St. David's.] Dr. Hicks' cabinet.
		Agnostus cambrensis, Hicks. Allied to *A. integer* of Bohemia.	[St. David's.] do.
		Agnostus acumensis, Hicks, and **A. lagena** are new forms.	[St. David's.] do.
Gh		**Agnostus scutalis,** Salter. Resembles *A. pisiformis,* Linn. from Sweden.	St. David's. do.
Gh		**Agnostus Barrandii,** Hicks. Smooth, convex, head-lobes incomplete. Resembles *A. lævigatus* of Sweden.	b. 265, b. 274, Lower beds, St. David's. (D. Homfray.)
		Agnostus Eskriggi, Hicks.	[St. David's.] Dr. Hicks' cabinet.
Gh		**Agnostus reticulatus,** Ang. Pal. Suec. The limb is deeply and largely reticulated. (In some collections this appears as *A. nodosus,* Belt.)	b. 267, b. 272, Tyn-y-groes. on the Mawddach River. (Mr. Homfray.)
Gh	*Microdiscus.*	**Microdiscus punctatus,** Salter (Quart. Geol. Journ. Vol. xx. pl. 13, fig. 11). A minute form, shewing 4 body segments. Otherwise it has the characters of *Agnostus,* and like it is blind, and has no facial suture.	St. David's. (Dr. Hicks.) a. 253, Tyddyngwladis. (Mr. Plant.) b. 275, Tafarn Helig, S. of Ffestiniog, N. Wales. (Mr. D. Homfray.)
Gh		**Holocephalina primordialis,** Salter (ib. fig. 9). A very simple type of many-ringed trilobites allied to *Arionellus,* which is a Bohemian and Spanish form. It has minute external eyes, and the facial suture only cuts off the extreme spinous angle.	a. 288, St. David's. (Pres. Dr. Hicks; and cast of figured specimen, by J. W. Salter.) b. 304, Maentwrog waterfall valley, or Tafarn Helig, Ffestiniog. (Pres. by D. Homfray.)

Case and Column of Drawers.	Reference to McCoy's Synopsis: and Figures of Genera.	Names and References; Observations, &c.	Numbers and Localities.
G h	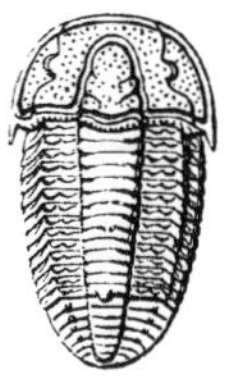	**Conocoryphe variolaris,** Salter (ib. figs. 6, 7). *Conocoryphe* (or *Conocephalus,* as it is often written) belongs to a group of trilobites intermediate between *Calymene* and *Olenus* (see Bala Group), and is extremely common in the Lingula flags, and the older Cambrian rocks in all countries. (*Ptychoparia,* Corda.)	St. David's. (Dr. Hicks.)
G h	*Fry of Conocoryphe,* greatly magnified.	**Conocoryphe applanata,** Salter (Quart. Geol. Journ. Vol. XXV. pl. 2, figs. 1, 2, 4; 1869). A small species, with flattened body.	St. David's. (Presented by Dr. Hicks.)
G h	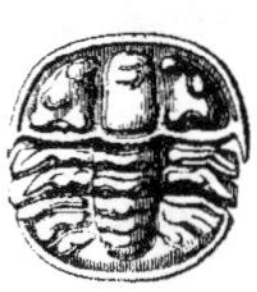	**Conocoryphe humerosa,** Salter. Convex, and with spines to the pleuræ. (Vol. XXV. ib. pl. 2, fig. 7.)	Cast only. Do. Do.
G h	The fry of all Trilobites possesses fewer segments, often but one or two, and is blind, and without suture.	**Conocoryphe bufo,** Hicks (Quart. Geol. Journ. 1869, pl. 2, fig. 8). A massive species, granulate all over, with a great produced front.	a. 476, cast, (pres. by Dr. Hicks,) lowest beds of Menevian Group, St. David's.
G h		**Conocoryphe? Homfrayi,** n. sp. A fine species $3\frac{1}{2}$ inches long, with slightly deflected pointed pleuræ.	b. 350, Maentwrog Valley, base of Menevian Group. (D. Homfray.)
G h		**Conocoryphe?** 2 sp. Both these species are much like Olenus, but the pleuræ are facetted and deflected as in *Conocoryphe.*	b. 270, Tafarn Helig. near Trawsfynydd. (D. Homfray.) b. 355, Waterfall valley, base of Menevian. (D. H.)
G h	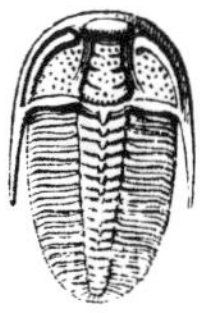	**Conocoryphe coronata,** Barr. Boh. Syst. Silur. pl. 13, fig. 20. Prof. Corda separated this from the other species of *Conocephalus,* on account of the absence of eyes. But it seems that Zenker intended this group for the true Conocephalus: hence we must separate the others, rather than this.	b. 307, Head of the Rhaiadr ddu valley, base of the formation. This locality is very rich in Trilobites. Nearly all the St. David's species are found there. (Pres. by D. Homfray.)
G h	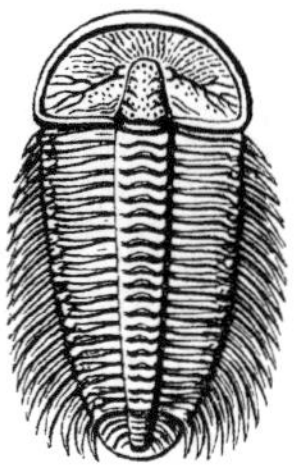	**Erinnys venulosa,** Salter (Quart. Geol. Journ. ined.). A trilobite with minute eyes, no facial suture, and the most numerous body-segments, twenty-three. It is possibly not distinct from *Harpides,* Beyrich. Two species are figured from Sweden by Angelin.	a. 484, St. David's. (Presented by Dr. Hicks.) b. 301, 302, Waterfall valley, S. of Maentwrog. b. 303, perfect—same locality. (Mr. D. Homfray.)

Case and Column of Drawers.	Reference to McCoy's Synopsis: and Figures of Genera.	Names and References; Observations, &c.	Numbers and Localities.
G case and Gh		**Paradoxides Davidis,** Salter (Quart. Geol. Journ. Vol. xx. pl. 13, figs. 1—3). The largest trilobite known in Britain: often 18 to 20 inches long. This fine species, discovered in 1863, belongs to a genus everywhere characteristic of the Menevian group only, and found in Bohemia, Spain, Sweden and Norway, Newfoundland, and the United States; in most cases it is associated with the preceding and following genera.	a. 2, a. 8, St. David's. (a. 2, Cast, presented by J. W. Salter, of the largest specimen known, in the Brit. Museum). a. 3, specimen shortened by the cleavage. b. 276—278. (Pres. D. Homfray). St. David's. a. 261, a. 262, The Gold mines, Dolgelly, N. Wales. Presented by Mr. Plant, 1869. a. 296 (Dr. Hicks).
Gh		[**P. Harlani,** Green. A cast is placed in the drawer for comparison. It is the only species known to attain greater size than *P. Davidis.* The species comes from Boston, Massachusetts].	a. 10.
Gh	*P. Davidis.*	**Paradoxides Hicksii,** Salter (Quart. Geol. Journ. Vol. xxv. pl. 3). The form and sculpture is quite intermediate between *Paradoxides* proper and the next genus. And it is worthy remark, that in geological position it is antecedent to *Anopolenus* next described, as that precedes *Olenus.*	a. 479, a. 480, St. David's, bottom beds. (Presented by Dr. Hicks. a. 480 is a specimen figured in the Quart. Geol. Journ.) a. 251, Camlan River, N. of Dolgelly. (Mr. Plant.)
		[**P. Aurora,** Salter. From still lower beds; in Dr. Hicks' cabinet.]	
Gh		**Anopolenus Henrici,** Salter (Quart. Geol. Journ. Vol. xx. pl. 13, figs. 4, 5, and Vol. xxi. p. 478). One of the most curious trilobites known, with immensely long eyes, contracted cheeks, great head-spines, and the hinder pleuræ dilated. *Parad. Loveni,* Angelin, is of this genus.	a. 481, St. David's, Cast of the figured specimen (Q. G. Journ.). b. 295, 296, Head of the Rhaiadr ddu valley. (Mr. D. Homfray. 296 is figured Q. G. Jour.) a. 249, Tyddyngwladis, Dolgelly. (Mr. Plant.)
Gh	*Anop. Salteri.*	**Anopolenus Salteri,** Hicks (Quart. Geol. Journ. ref. Vol. xxi. p. 478, f. 1). A much longer and narrower species than the last, with equally large eyes. The development in excess of the hinder pleuræ (or leg-bases) is in accordance with the affinity with *Paradoxides* (see last page).	St. David's. Base of the Menevian beds, alternating with top of Harlech grits. a. 477, young, and a. 478 adult. (Pres. by Dr. Hicks.) a. 250, Camlan River. (Mr. Plant.) b. 305, Rhaidr ddu valley. (Mr. D. Homfray.)

Case and Column of Drawers.	Reference to McCoy's Synopsis: and Figures of Genera.	Names and References; Observations, &c.	Numbers and Localities.
G h	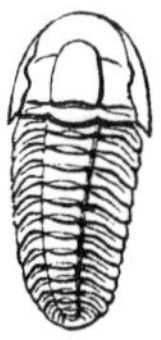	[**Arionellus longicephalus,** Hicks MSS. A genus new to Britain. In Dr. Hicks' cabinet.] The less convex and less distinct glabella, and the narrower axis of the body, distinguish this, as a species, from *A. ceticephalus,* Barr.	St. David's. Cast, presented by Dr. Hicks.
		[It is possible the two species of *Olenus* and the *Hymenocaris major* should be included in the Ffestiniog Group. They lie close under it.]	
		Olenus, Dalman. With many of the characters of *Paradoxides,* but with a glabella narrowed in front, and but 14 body rings at most.	
G h	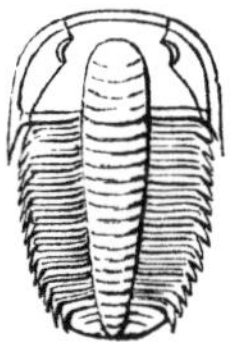	**Olenus** ———, sp. (*Olenus gibbosus,* Belt. Geol. Mag. 1867, not of Swedish authors (?). The Swedish species are not sufficiently figured yet: and ours appears to have too many spines to the tail, and a wider front.	b. 268, b. 269, River Mawddach, at Tyn-y-groes, N. Wales. (Mr. D. Homfray.)
G case and G h		**Olenus cataractes,** Salter (Decades Geol. Survey, No. XI. pl. 10, fig. ref.). Differs from the common species in the next formation, by the gradually tapering, not abruptly narrowed, body. Mem. Geol. Surv. III. p. 300. The specimens b. 294, pres. by Mr. D. Homfray, shew the effects of cleavage in shortening or lengthening the body, the spines, &c. In this state it might be taken for many different species.	a. 7, Maentwrog falls, Ffestiniog. b. 294, Caen-y-coed, near the falls. (Mr. D. Homfray.) a. 11, Ynys cynhauarn, Treflys, near Portmadoc. (Pres. by Mr. Ash.) a. 246, a. 248, Cae Gwernog, Upp. Mawddach: and a. 247, Cefn Deiddwr. a. 261, Moel Ispray. (All pres. by Mr. Plant.)
	Phyllopods. The small bivalve species have their affinity with the *Limnadia,* but look like water-fleas (*Cypris*).	**Primitia Solvensis,** R. Jones, Ann. N. H. s. 4. 2. 55. Specimens in Mus. Pract. Geology.	St. David's. (Dr. Hicks.)
G h		**Primitia buprestis,** Salter (*punctatissima,* Jones) A large species like a small bean.	a. 287, do. do.
G h		**Primitia vexata,** Hicks. Brit. Assoc. Report.	a. 282, do. do.
G h	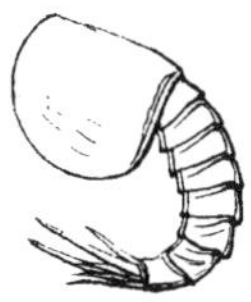	**Hymenocaris** (*Saccocaris,* Halifax Trans. 1867) **major,** Salter, n. s. A large ovate carapace, strongly emarginate behind, and larger than *H. vermicauda,* see p. 10. Body segments broad and short, at least in seven of the anterior ones. Appendages not known.	b. 297, Caen-y-coed, near Maentwrog. (Mr. D. Homfray.) b. 297, body segments of the same. Same locality and donor.

Case and Column of Drawers.	Reference to McCoy's Synopsis: and Figures of Genera.	Names and References; Observations, &c.	Numbers and Localities.
	Brachiopods.	**MOLLUSCA Brachiopoda.**	
	Bivalves protected by upper and lower shells, not by side-plates as in ordinary bivalves.	**Discina? pileolus** sp. Salter. Minute cap-shaped horny Brachiopods are found in all the Paleozoic deposits. (For genus see Bala group).	St. David's. (Presented by Dr. Hicks, 1869.)
G h		**Obolella? sagittalis,** Salter (Davidson, earliest Brach. Geol. Mag. 1868, Vol. v. pl. 15, figs. 17—24). The genus not quite certain, but *Obolella* is common in old rocks.	St. David's. (Presented by Dr. Hicks, 1869.)
G h		**Obolella nucleata,** MSS. A very small apiculate species.	a. 257, Gwynfynydd, Gold mines. a. 258, Cwm heisian. (Mr. Plant, 1869.)
G h	Brachiopods have spiral arms, which are probably the labial tentacles. The respiration is carried on, not by gills, but by the mantle-surface. Brachiopods (the horny kinds) precede the ordinary bivalves in time.	**Obolella maculata,** Hicks, ib. pl. 16, figs. 1—3.	St. David's. (Dr. Hicks.)
		Lingulella ferruginea, Salter (ib. pl. xv. figs. 1—4). The earliest *Lingulæ* differed from modern ones in having a groove under the beak for the passage of the pedicle, and thus being more like *Obolus.* Quart. Geol. Journ. 1867, Vol. xxiii. p. 340, fig. 1. (See p. 2.)	St David's. (Dr. Hicks.)
G h		**Orthis Hicksii,** n. sp., Davidson (ib. pl. 16, fig. 17). Very like *O. Carausii,* Salter, but with a short hinge. (For genus, see Arenig Group.)	a. 483, in sandstones at Porth Rhaw, St. David's. (Dr. Hicks.)
	Univalves.	**MOLLUSCA Pteropoda.**	
G	(Sea Butterflies.) The Pteropods are greatly developed in the very oldest rocks; and are low forms of mollusks, floating and almost flying in deep water by means of the double-lobed foot. The breathing cavity is a ciliated surface. The operculum has been discovered in many species.	**Theca obtusa,** Salter (Mem. Geol. Surv. Vol. iii. p. 352, fig. 17). A large species of a genus like the much smaller living *Creseis.*	a. 4, Maentwrog falls, N. Wales. Mr. Ash. (Upper beds of the formation.)
G h		**Theca corrugata,** Salter (Quart. Geol. Journ. Vol. xx. pl. 13, fig. 10). A moderately broad species, with rugose lines of growth.	St. David's. (Pres. Dr. Hicks.) b. 264, Rhaidr ddu valley, Maentwrog. (Mr. D. Homfray.) a. 259, a. 260, near Gold mines, Mawddach River. (Mr. Plant.)
		Theca Homfrayi, Salter, n. s. Much longer than *T. corrugata.*	b. 263, Tyddyngwladis, gold mines. (D. Homfray.)
		Theca penultima, Salter. [Dr. Hicks' cabinet.]	
		Theca stiletto, Hicks. Brit. Assoc. Repts.	St. David's. (Dr. Hicks.)
		Stenotheca cornucopiæ, Salter, new genus.	a. 279, do. do.
		Cyrtotheca hamula, Hicks. Like Theca, but strongly curved.	a. 280, do. do.

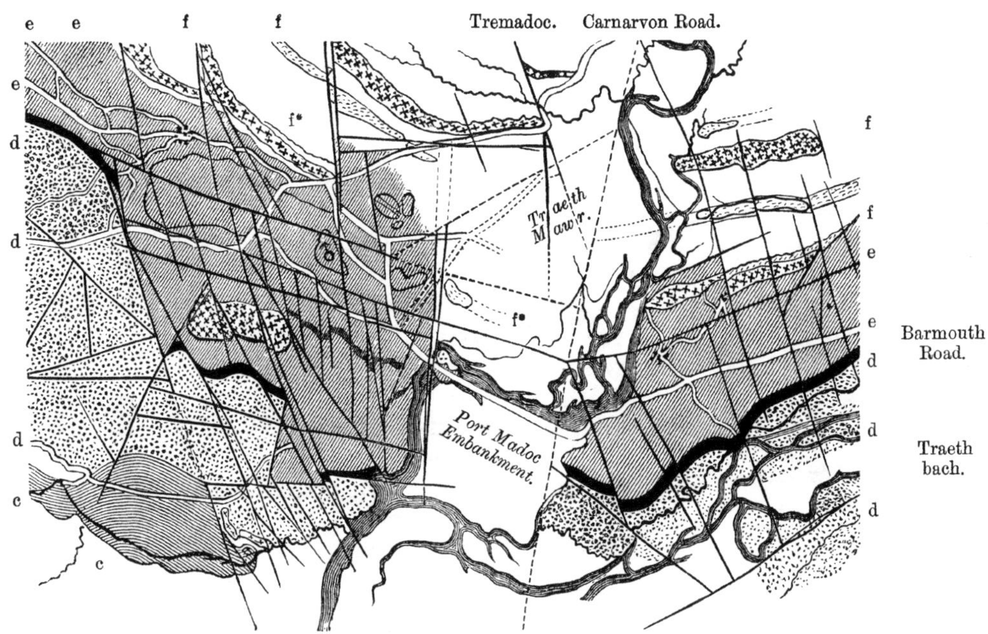

PORTMADOC ESTUARY: shewing the relations of the Menevian (c), Ffestiniog (d), Tremadoc (e), and Arenig Groups (f) in the district around Portmadoc, N. Wales. By the Rev. A. Sedgwick, LL.D. 1847, and J. W. Salter, 1853—7. The faults are all by J. W. S. The strata, in descending order, are:—

f. Arenig (or Skiddaw) Group. Dark earthy slates on a base of sandstone (=Stiper Stones, Shropshire).
e. Tremadoc Group. Dark slates, iron stained, and with felspathic beds—1500 feet thick.
d. Ffestiniog Group. Thick flaky sandstones, 2000 feet, and a bed of *black* slate 300 feet.
c. Menevian Group. Dark slate and sandstone—only the top beds (with *Olenus cataractes*).

d. Ffestiniog Group { Middle / Upper } Lingula flag. { Sandstones, hard, laminated, flinty, 2000 feet thick. / Black Slates, often stained with iron, 200 to 1000 feet.

Case and Column of Drawers.	Reference to McCoy's Synopsis: and Figures of Genera.	Names and References; Observations, &c.	Numbers and Localities.
		AMORPHOZOA. None yet known.	
	Graptolites. Horny sheaths, with a slender solid axis, and with numerous close-set cells opening into the common sheath. The mouths often armed with spines.	**POLYZOA or HYDROZOA?** (*Graptolitidæ*). Considered by Prof. McCoy and some other authors as *Sertularian Zoophytes.* I therefore place them here in the order given to them in the Synopsis. (See for their affinities, Mem. Geol. Surv. Vol. III. p. 328, and Carruthers, Geol. Mag. Vol. V. pl. V.)	
Gh	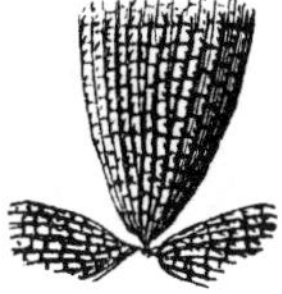	**Dictyonema (Graptopora) sociale,** Salter (Mem. Geol. Surv. Vol. III. pl. 4, fig. 1). Variously known as *Gorgonia, Dictyonema,* and *Graptopora.* A net-like fossil, allied to the *Fenestella* and *Polypora* of the Bala and Wenlock rocks: but of horny texture and with cells in double row, like Graptolites. (Salter, Mem. Geol. Surv. Vol. III. p. 331.)	a. 12, black slates of Tremadoc and Wern, and other places, Cefn Cyfarnedd &c. on the Carnarvon Road. (Presented by Mr. Ash, 1862.) a. 13, olive shales of White-leaved oak, Malvern. (Rev. W. Symonds.)

Case and Column of Drawers.	Reference to McCoy's Synopsis: and Figures of Genera.	Names and References; Observations, &c.	Numbers and Localities.
	Worm-tubes and burrows.	**ANNELIDA** (The burrows are frequently called Fucoids, see note in Wenlock Group).	[These must be obtained for the collection.]
	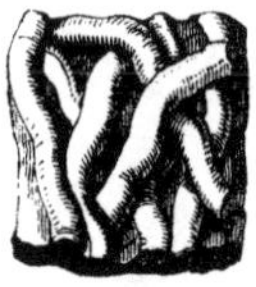	**Scolecites** (Irregular worm-burrows in the sandy silt, crossing each other at various angles. Such burrows, filled with the excreta of the worm, are common in all rocks. (Salter, Quart. Geol. Journ. Vol. IV. p. 223).	[Near Bangor: Ffestiniog, Dolgelly, &c. common.]
G h	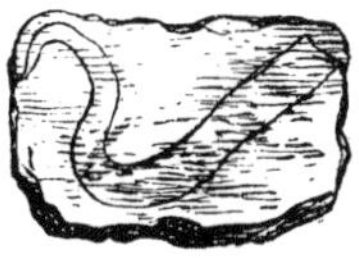	**Scolecoderma antiquissimum,** Salter (*Trachyderma* in various works) is a common large worm-tube in the Hollybush sandstone, which sandstone I have referred to the middle Lingula flags. *Scolecoderma* is the tube itself, often membranous, of the worm.	Hollybush sandstone, Malvern. (Presented by Dr. Grindrod.)
		**Cruziana semiplicata,** Salter (Mem. Geol. Surv. III. pl. 3). A curiously plaited tube, furrowed along the middle. The genus was very abundantly spread over the world in Cambrian times. It is rather an anomalous form. But the ridgy plaiting of the upper surface probably has reference to the double rows of lamellæ covering the back of the short broad worm.	[Common near Bangor.]
	Phyllopod Crustacea, allied to *Apus* and *Nebalia.*	**CRUSTACEA. Phyllopoda.**	
G case and G h		**Hymenocaris vermicauda,** Salter (Mem. Geol. Surv. Vol. III. 1866, pl. 1, fig. 2, p. 293). A shrimp-like form, evidently allied to the living *Nebalia* and the Silurian *Ceratiocaris.* The carapace is ample and undivided. Body with 7 or 8? joints. Tail with 6 appendages.	a. 15, Borth harbour, Portmadoc. (Presented by Mr. Salter.) Top beds of the middle Lingula flag. a. 16 (Pres. by Mr. F. Ash). b. 16 (Mr. D. Homfray).
	Trilobites.	**CRUSTACEA Trilobita.**	
G h		**Agnostus trisectus,** Salter, Decade 11, M. G. S. pl. 1, fig. 11. The centre tail-lobe is triple.	a. 22, Whiteleaved oak, Malvern. Black Shales.
G h	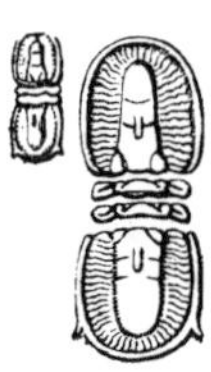	**Agnostus princeps,** Salter (Decade 11, l. c. figs. 1—5); also Mem. Geol. Surv. Vol. III. pl. 5, fig. 1 a). (*A. princeps,* var. Salter, Mem. Geol. Surv. l. c. pl. 5, fig. 16). A common Lingula flag species, with radiating striæ to the limb of the head.	Carreg wen, Portmadoc. (Mr. Ash and Mr. Homfray.) a. 267, a. 268, Moel Gron, Upper Mawddach. (Mr. Plant, 1869.)

Case and Column of Drawers.	Reference to McCoy's Synopsis: and Figures of Genera.	Names and References; Observations, &c.	Localities and Numbers.
G h		**Agnostus,** sp. with narrow glabella.	Carreg wen. (Mr. Homfray.)
G h		**Agnostus,** sp. undetermined.	a. 269, do. (Mr. Plant.)
G h		**Olenus micrurus,** Salter (Mem. Geol. Surv. Vol. III. 1866, pl. 2, figs. 5, 6. p. 300). Distinct by the body being attenuated behind, and by the small entire tail of two segments.	a. 263, Penmaen Pool, Mawddach valley, near Dolgelly, middle Ling. flag. (Mr. Plant.)
G h		**Olenus Plantii,** Salter, n. sp. An oval species much flattened and expanded. Mr. J. Plant, who discovered it, has distributed casts and photographs of this fine and well marked fossil.	Casts from Craig-y-Dinas; and a. 272, Moel Gron, Upper Mawddach. (In Upper Ling. flag. Mr. Plant, 1869.)
G h	*Typical Olenus.*	**Olenus,** small sp. like *O. cataractes.* See page 6.	a. 276. Penmaen Pool, Mawddach R. (In Upper L. flag.) Mr. Plant.
G h		**Olenus (Parabolina) spinulosus,** Wahl. (Angelin. Pal. Suec. t. XXV. fig. 9). Only lately noticed in Britain. It is common, and known by the fringe of long spines, directed backwards, to all the body segments. *O. serratus,* Salter (Mem. Geol. Surv. III. pl. 5, figs. 6, 7) is the same species. (*O. comatus,* Plant MSS.)	a. 261, Penmaen Pool, near Dolgelly (Mr. Plant). a. 260, 271, 277, Rhiw felyn, on the Upper Mawddach R. (Mr. Plant). b. 349, fine series, same locality. (Mr. D. Homfray.)
G h		**Olenus (Peltura) scarabæoides,** Wahl. *Peltura* is a natural subgenus, the opposite of *Sphærophthalmus,* in having the axis very broad, the cheeks small, and the head spines reduced to zero. Decade 11, Geol. Surv. pl. 8.	b. 285, Rhiw felyn, Upper Mawddach. (Mr. D. Homfray.) a. 270, Moel Gron, Upper Mawddach. (Mr. Plant.) Malvern Black Shales. (Rev. W. Symonds.) All in Upper Lingula flag.
G h	*Sphærophthalmus.*	**Olenus (Sphærophthalmus,** Ang.) **flagellifer,** Angelin (Salter in Decade 11, Geol. Surv. pl. 8, figs. 7, 8) *O. flagellifer* and *O. alatus, pecten, humilis, bisulcatus,* are members of the subgenus *Sphærophthalmus,* in which the head is wide and short, and the globular eyes remote. *Sphærophthalmus* is probably a very distinct genus.	Carreg wen, Borth, Portmadoc. (Presented by Mr. Ash.) *Black slates* of Upper Lingula flag. b. 283, same loc. Mr. D. Homfray (fig. in Decade 11).

Case and Column of Drawers.	Reference to McCoy's Synopsis: and Figures of Genera.	Names and References; Observations, &c.	Localities and Numbers.
G h		**Olenus (Sphær.)** sp.	Carreg wen, Borth. (Mr. D. Homfray.)
G h		**Olenus (Sphær.) alatus,** Bœck (ib. pl. 8, fig. 6). Head spines remarkably curved in this species.	Carreg wen, Borth. (Mr. Ash.) Malvern, Mr. Symonds.
G h	The typical number of body segments in the higher forms is 10—13, but these genera have usually more, and the primordial trilobites are generally of extravagant forms, or else of very simple structure.	**Olenus (Sphær.) expansus,** n. sp. A very narrow axis and enormously wide flanks distinguish this.	a. 275, Moel Gron. (U. Ling. flag.) Mr. Plant.
G h		**Olenus (Sphær.) humilis,** Phillips (ib. pl. 8, figs. 9, 10, 11). Only 7 body rings. The most minute species known. Extremely common at Malvern.	a. 273, Moel Gron, Upper Mawddach. (Mr. Plant.) a. 23, Malvern Black Shales. (Mr. Symonds.)
G h		**Olenus (Sphær.)** sp.	Malvern. (Mr. Symonds.)
G h		**Olenus (Sphær.) bisulcatus,** Phillips (ib. pl. 8, fig. 6). The body and tail of this species, are known, but are not in this collection.	Malvern. Rev. W. Symonds.
G h	*O. Pecten.*	**Olenus (Sphær.)** sp. like *O. bisulcatus.*	b. 281, Rhiw felyn, Upper Mawddach. (Mr. D. Homfray.)
G h		**Olenus (Sphær.) pecten,** Salter (ib. pl. 8, fig. 12). The most spinose and abnormal of all the British species. The cheeks are singularly contracted: the tail 13-spined (Quart. Geol. Journ. XXI. p. 478).	b. 308, Malvern. (Rev. W. Symonds.)
G h		**Olenus (Sphær.)** sp.	Carreg wen, Borth. (Mr. Homfray.)
G h		**Conocoryphe? ecorne,** Angelin? (*Acerocare ec.* Pal. Suec. XXV. fig. 10). If this be Angelin's species, which seems likely, *Acerocare* is a needless name.	b. 259. Penmaen Pool, Dolgelly (Middle Lingula flags. Mr. Plant).
G h		**Conocoryphe Williamsoni,** n. sp. A tumid front, spinose cheeks, and pointed bent pleuræ, distinguish this.	b. 292, Rhiw felyn. Upper Mawddach. (Mr. D. Homfray.) b. 258, Penmaen Pool, Dolgelly. (Cast) Mr. Plant.

Case and Column of Drawers.	Reference to McCoy's Synopsis: and Figures of Genera.	Names and References; Observations, &c.	Localities and Numbers.
G h		**Conocoryphe (Conocephalus) invita,** Salter (Decade XI. pl. 7, fig. 6). This genus is one of those which unite the *Oleni* of the Middle Cambrian with the *Calymenidæ* of the Upper Cambrian and Silurian beds. 14 body rings.	Penmorfa church, Tremadoc. Upper Lingula flags. b. 290, 291, Carreg wen, Borth. [Mr. D. Homfray.]
G h		**Conocoryphe abdita,** Salter, Mem. Geol. Surv. Vol. III. pl. 5, figs. 13 and 15.	b. 284, Ogof ddu, as below. (Mr. D. Homfray.)
G h		**Dikellocephalus celticus,** (Salter, Mem. Geol. Surv. Vol. III. pl. 5, fig. 22). A genus related to *Olenus*, and yet differing from it by the expanded form and large tail. Probably this form and *D. furca* of the Tremadoc Group belong to the genus *Centropleura* of Angelin. Several British and Swedish species are known.	Craig Ogof ddu, near Criccieth; in upper Lingula flag. (Mr. Ash.) b. 282, specimens figured in Mem. Geol. Soc. Same locality. [Mr. D. Homfray.]
G h		**Dikellocephalus,** sp.	Carreg wen, Portmadoc. (Mr. Ash.)
	Brachiopods, or Lamp-shells.	**MOLLUSCA BRACHIOPODA.** Almost all the species known of this order in the primordial group are the horny species, such as *Discina, Lingula, Lingulella, Obolella,* &c. One or two species of *Orthis* are all the calcareous hinge-bearing forms known.	
G h		**Lingulella lepis,** Salter (Mem. Geol. Surv. Vol. III. p. 334, fig. 11. Davids. Sil. Brach. pl. 3, figs. 53—59). *Lingulella* differs from the modern *Lingula* by having a groove for the passage of the pedicle, else it is quite like it.	Ogof ddu, Criccieth. (Mr. Ash.) b. 312, Dolgelly.
G case and G h	As *Lingula Davisii,* Pl. 1 L, fig. 7; as *L. ovata,* Pl. 1 L, fig. 6, pp. 252, 254. (Note. Not the original *L. ovata*—that name is retained for the Bala species, p. 254). And when distorted, as *Tellinomya lingulæcomes,* Pl. 1 K, fig. 18.	**Lingulella Davisii,** McCoy (Davids. Sil. Brach. t. 4, figs. 1—16). The common Brachiopod of the Lingula flag: of all shapes according to the pressure of the rock. Its true shape is satchel-shaped (McCoy). It occurs both in the sandy middle portion of the Lingula flag, and in similar strata in the Tremadoc rocks. It occurs of smaller size in more slaty deposits, but its place in these is usually taken by the *L. lepis.*	Borth, Portmadoc. (Mr. Ash and Mr Homfray.) a. 19, Said to be from E. of Nant-y-groes? a. 17, Penmorfa, Portmadoc (as *Tellinomya*). a. 20 (As *L. ovata*), Penmorfa. a. 264, Hafod Owen, Upper Mawddach R. (Mr. Plant.)

Case and Column of Drawers.	Reference to McCoy's Synopsis: and Figures of Genera.	Names and References; Observations, &c.	Localities and Numbers.
G h		**Lingulella** ———, sp.	Dolgelly. (Mr. D. Homfray.)
G h		**Kutorgina cingulata,** Billings (Davids. Sil. Brach. pl. 4, figs. 17—19). *Obolella Phillipsi,* Holl. (Quart. Geol. Journ. Vol. XXI. p. 10, fig. 10 a, b). A very unusual shape for the genus, one valve nearly flat, and with a straight hinge-line: the smaller valve convex.	b. 308, 310, Ogof ddu, Criccieth. (D. H.) a. 18, Grits on Hollybush Hill, Malvern.
G h		**Obolella Salteri,** Holl. (l. c. fig. 9). A small and rounded species, like those of Canada (Davids. Sil. Brach. t. 4, figs. 28, 29).	a. 21, Black Shale, Malvern. (Rev. W. Symonds.)
G h		**Obolella** ———, sp.	b. 309, Dolgelly, Penmorfa church. Upper Lingula Beds. (D. H.)
G case G h		**Orthis lenticularis,** Dalman (Salter, Mem. Geol. Surv. Vol. III. pl. 4, figs. 8—10). A small shell seldom half an inch wide, and rather wider than long. The valves are both gently convex. The number of principal ribs 10 or 12, interlined by smaller ones, and all crossed by interrupted and rather wavy ridges of growth, so that the surface is somewhat reticulated (*Atrypa lenticularis,* Dalman, *Spirifer lentic. Von Buch.* Mr. Davidson has very fully figured this common Upper Lingula flag species in his Monograph of the Cambrian and Silurian Brachiopods. Palæont. Transactions, Pl. XXXIII. fig. 22, &c.).	a 265, Craig-y-dinas, Upper Mawddach. a. 266, Rhiw felyn; do. (Mr. Plant.) A common shell in all our Upper Lingula flag localities: in N. Wales, viz. Tremadoc, Criccieth, Dolgelly gold mines, &c. Ogof ddu, Criccieth. (Mr. Homfray.)

MIDDLE CAMBRIAN. *e.* Tremadoc Group. N. Wales. S. Wales. (Sedgwick, 1847).

e 1. *Lower.* Chiefly Black Slate.

e 2. *Upper.* Sandstones, grey and bluish: iron beds: ferruginous slate, &c.

e 1. LOWER TREMADOC SLATE (Salter, 1857). A natural continuation of the Upper Lingula Flags. Chiefly black slate, but very ochreous in part. The fauna is essentially Middle Cambrian, but shews a tendency to include some few Lower Bala types, such as *Niobe, Psilocephalus,* &c. The species however are all distinct, even from those of the Arenig Group, and those of the Upper are distinct from those of the Lower Tremadoc. We distinguish the upper, *e* 2, from the lower group, *e* 1.

Case and Column of Drawers.	Reference to McCoy's Synopsis: and Figures of Genera.	Names and References; Observations, &c.	Localities and Numbers.
G case and Gi	*Trilobites.*	**Niobe Homfrayi,** Salter (Pal. Soc. Trans. Vol. I. pl. 20, fig. 3). A genus intermediate in some sort between *Asaphus* and *Ogygia.* The labrum is round-pointed as in *Ogygia,* and the glabella lobed as in that genus, but broad as in *Asaphus.* The pleuræ bent, the axis of the body broad, the tail ample.	Tyn-y-llan, Tyn-y-dre, and other places near Tremadoc and Portmadoc. (Mr. Ash.) b. 329, 330, 362, (Mr. D. Homfray's fine specimens figured in the Pal. Tr.) from Penmorfa church.
Gi		**Psilocephalus innotatus,** Salter (id. pl. 20, figs. 13, 14). A small Trilobite, allied to *Illænus,* very abundant.	Tyn-y-llan, Tyn-y-dre, and other places near Borth, Portmadoc. (Mr. Ash.) b. 352, 353, figured specimens. (Mr. D. Homfray.)
Gi		**Psilocephalus inflatus,** Salter (Mem. Geol. Surv. Vol. III. p. 316, woodcut 8).	b. 357, Penmorfa. (Mr. Homfray). Figured sp.
Gi		**Conocoryphe depressa,** Salter (Mem. Geol. Surv. Vol. III. pl. 6, figs. 1—3). The species of this genus which occur in rocks above the Lingula flag, differ in not having the glabella lobes pronounced.	Do. (Mr. Homfray).
Gi		**Conocoryphe verisimilis,** Salter (Mem. Geol. Surv. Vol. III. pl. 6, fig. 13).	b. 354, Penmorfa village. (Mr. Homfray.)
Gi		**Dikellocephalus (Centropleura) furca,** Salter (Mem. Geol. Surv. Vol. III. pl. 6, fig. 4, and pl. 8, fig. 10). The two blunt prongs to the hinder part of the tail distinguish this from *D. celticus.* It has also fewer ribs.	Llanerch, Moel-y-gest. Garth, Portmadoc. (Mr. D. Homfray.)

Case and Column of Drawers.	Reference to McCoy's Synopsis: and Figures of Genera.	Names and References; Observations, &c.	Localities and Numbers.
	Pteropods.	**PTEROPODA** (Sea-Butterflies).	
G i		**Theca operculata,** Salter (Mem. Geol. Surv. Vol. III. pl. 10, figs. 22—24). Most of the species of fossil Theca are operculated. The living Pteropods nearest them in shape are not so, but several of the spiral kinds have an operculum. (ib. p. 351 and note).	Tyn-y-llan, &c., near Borth, Portmadoc. (Mr. Ash.) b. 335. (Pres. Mr. Homfray.)
G i		**Theca bijugosa,** Salter (ib. pl. 10, figs. 19, 20). The species differ in proportionate length and breadth, and in the longitudinal ridges; most are smooth.	Borthwood, Portmadoc. (Mr. Ash.) b. 334, Tyn-y-dre, do. figured specimen. (Mr. Homfray.)
		Theca arata, Salter (ib. pl. 10, figs. 15 and 21). May easily be obtained.	[Portmadoc.]
	Brachiopods.	**BRACHIOPODA.**	
G i		**Lingulella lepis,** Salter (Davids. Sil. Brach. in Pal. Trans. t. 3, figs. 53—59).	Tremadoc. (Mr. Homfray.)
	Trilobites and Phyllopods.	*e* 2. Upper Tremadoc Slate (Salter, 1857).	
G i		**CRUSTACEA (Phyllopods and Trilobites).**	
		Lingulocaris lingulæcomes, Salter (Mem. Geol. Surv. Vol. III. pl. 10, figs. 1. 2). A bivalve crustacean, allied to the *Ceratiocaris* of the Silurian rocks, and also to the living *Limnadia.*	a. 28, Garth, Portmadoc. (Mr. Ash.) b. 327, figured specimen. (Pres. Mr. Homfray.)
G i		**Ceratiocaris? latus,** Salter; front segments of body very broad. Badly figured in p. 294, Mem. Geol. Surv. Vol. III. as *Hymenocaris.* For genus, see Ludlow Rocks.	b. 299, Garth (figured specimen, Mr. D. Homfray).
		Ceratiocaris ? insperatus, Salter (Mem. Geol. Surv. Vol. III. p. 295, fig. 6). It is quite probably the tail portion of the preceding.	b. 343, Above Penmorfa Railway cutting, figured specimen. (Mr. Homfray.)

Case and Column of Drawers.	Reference to McCoy's Synopsis: and Figures of Genera.	Names and References; Observations, &c.	Numbers and Localities.
		TRILOBITA.	
G i G case		**Angelina Sedgwickii,** Salter (Decade Geol. Surv. 11. pl. 7). The most abundant of the Tremadoc trilobites: allied to *Olenus*, but with 15 body rings (*Olenus* has 14), and greatly larger than any known species of that genus. Even size is a character of some importance in classification. Occurs in all sorts of compressed shapes in the slate and sandstone.	a. 32, Garth and Penclogwyn, at Portmadoc; Carnarvon. (Mr. Ash.) a. 24, Reduced in length by cleavage. b. 337—340, Various ages of this fine species. (Mr. Homfray.)
G case		**Dicellocephalus furca,** Salter (Mem. Geol. Surv. Vol. III. pl. 8, fig. 9, 10). Belongs to the section *Centropleura* (ib. p. 303).	a. 30 (Mr. Ash.) Moel-y-gest; in the bottom beds of the Upper Tremadoc. a. 30*, Garth. (Mr. Homfray.)
G case	*Isotelus.* Pl. 1 F, fig. 3, p. 169.	**Asaphus (Isotelus?) affinis,** McCoy. Perhaps distinct from the following species. The facial suture is marginal in front.	a. 26, Tremadoc, over Iron Works (Sedgwick). N.W. of Portmadoc.
G case		**Asaphus (Isot.?) Homfrayi,** Salter (Mem. Geol. Surv. Vol. III. p. 311, pl. 8, fig. 11—14). The earliest known form of this genus in Britain, if indeed the entire form of the labrum do not prove it a distinct genus nearer *Ogygia*. (See Lower Bala.) b. 344 is the labrum. The subgenus *Isotelus* is uncertain here, for the labrum is not notched as in that form.	a. 25, Garth and Penclogwyn, Portmadoc. (Mr. Ash.) b. 342, 344, 358, figured specimens in Salter's Mon. Brit. Tril. pl. 24. (Mr. Homfray.)
G i	*Labrum of Ogygia.*	**Ogygia scutatrix,** Salter (Mem. Geol. Surv. Vol. III. p. 312, pl. 8, f. 8, pl. 9, fig. 1). 8 or 9 inches in diameter, and nearly round. More like the French species from Brittany, *O. Desmaresti.*	Penclogwyn, Portmadoc. (Mr. Ash.)
G G i		**Cheirurus Frederici,** Salter (Mem. Geol. Surv. Vol. III. p. 322, pl. 8, figs. 1—3). A species with a general resemblance to the *C.* (*Eccoptochile*) *Sedgwickii,* McCoy (see Lower Bala), but with spines to the hinder rings of the body.	Garth, Portmadoc. (Mr. Ash.) b. 345, 346, 347. Mr. Homfray's specimen is figured in Mem. Geol. Surv. Vol. III. p. 323, fig. 10.
		PTEROPODA AND HETEROPODA.	
G i	*Univalve Shells. Pteropods.*	**Theca simplex,** Salter? (Mem. Geol. Surv. Vol. III. p. 352, pl. 11 B, fig. 22—26). This, if the same as the Arenig group species, is the only fossil common to both formations. In Canada there may be a transition from the Primordial Group to the Upper Cambrian, or at least to the Arenig Group: but not in Britain.	Garth, Tremadoc. (Mr. Ash.)
G case		**Theca sulcata,** n.s. Broad and short; with longitudinal folds.	b. 322, Llanerch, W. of Portmadoc, base of Upper Tremadoc. (Mr. Homfray.)

Case and Column of Drawers.	Reference to McCoy's Synopsis: and Figures of Genera.	Names and References; Observations, &c.	Numbers and Localities.
G case		**Theca (Centrotheca) cuspidata,** Salter (Mem. Geol. Surv. Vol. III. pl. 10, fig. 25). A long horn projects on each side from the mouth.	b. 336, Portmadoc. (Mr. Homfray.)
G case		**Theca trilineata,** MSS. A small species with longitudinal lines.	b. 333, Moel-y-gest, base of Upper Tremadoc. (Mr. Homfray.)
G case Gi		**Conularia Homfrayi,** Salter (Mem. Geol. Surv. Vol. III. pl. 10, figs. 11—13). A very large species, so slightly calcareous as to appear mere membrane on the slate. The puckering of the surface is probably due to unequal contraction. See Lower Ludlow Rocks.	Garth, Portmadoc. (Mr. Ash.) b. 323, 324, 325, figured specimens in above work. (Mr. Homfray.)
G case		**Bellerophon arfonensis,** Salter (l. c. p. 349, pl. 10, figs. 6—8). Squamose lines of growth, remote and rather regular.	Garth, Portmadoc. b. 328, 341 (Mr. Homfray, figured specimens).
Gi		**Bellerophon multistriatus,** Salter (l. c. p. 350, pl. 10, figs. 9, 10). With close decussating striæ.	Garth, Portmadoc. (Mr. Homfray.)
		CEPHALOPODA.	
G case and Gi		**Orthoceras sericeum,** Salter (l. c. Vol. III. p. 356, pl. 10, figs. 4, 5). This, and the *Cyrtoceras præcox* of the same work, are the oldest known forms of the Nautiloid, or shelly Cephalopod group. Cephalopoda became abundant in the succeeding period—the Bala group, even in its lower portion: and a few are known in the Arenig Group. b. 321 shews the septa.	Garth, Portmadoc. (Presented by Mr. Salter.) b. 321, 322, figured specimens in the work quoted. (Mr. Homfray.)

The great break, in organic life, between the 'Tremadoc Slate' and the 'Arenig or Skiddaw Group,' has disposed me, ever since I worked out their respective faunæ in the Tremadoc district in 1853, to regard the next overlying, or 'Arenig Group,' as the base of the great Upper Cambrian Group of Prof. Sedgwick.

Sir R. I. Murchison, in endeavouring to bring it first within the Ffestiniog or Lingula Flag Group, and later, among the Llandeilo or Lower Bala Group, has involved greatly the fossil evidence. But the Arenig or Skiddaw Group (Lower Llandeilo of Murchison) is peculiar, with the facies of the Lower Bala or Llandeilo Group, yet with *wholly distinct* species. To keep this Catalogue in harmony with the *Synopsis*, published in 1851—3, it is placed here as Prof. Phillips also regards it, as the terminal member of the Middle Cambrian.

The fossil evidence would permit us, with Lyell, to commence the Upper Cambrian (or Lower Silurian of Murchison) with this Group, which is well represented in the Stiper Stones district; though the fossils of that district were not described till 1859 by Murchison and myself, long after the Arenig Group, with its few fossils, found by Prof. Sedgwick in 1843, was established. The group was further illustrated by the fossils found in Skiddaw by Prof. Sedgwick, and described by McCoy, previous to 1851 (though the relative age of the rock was not then fully known). The right of nomenclature rests therefore with the Woodwardian Professor.

As the majority of the fossils, in both these transitional groups (Tremadoc and Arenig), have been first described by me, I may say that the Tremadoc Group seems to be the natural termination of the Ffestiniog or Middle Cambrian series; and the Arenig Group the true base of the Upper Cambrian (or Lower-Silurian of Sir R. I. Murchison). It is here treated as an intermediate Group. (J. W. S.)

MIDDLE CAMBRIAN. *f.* ARENIG or SKIDDAW GROUP (Sedgwick). 4000—5000 feet thick.

Base of Arenig and Arran Fowddy. The Stiper Stones Rocks. N.B. Some *few* of the Graptolites must be identified with those of true Lower Bala rocks, which overlie the Skiddaw Group proper in the Skiddaw district; but all the rest are distinct. The graptolitic or upper portion of the Quebec Group of Canada is identical with this: so are the graptolite gold-bearing shales of Victoria; and Prof. McCoy thinks several of the forms the same in each. I think the Victorian species are representatives only, but the genera are the same; while those of Canada are exactly ours in species and genera. The Angers slate is of this age. The whole group is unconformable upon the Tremadoc slate, and only one fossil is common to that formation.

Case and Column of Drawers.	Reference to McCoy's Synopsis: and Figures of Genera.	Names and References; Observations, &c.	Numbers and Localities.
	Graptolites, Simple, double-celled, and twin-graptolites; dichotomous, and bushy forms.	**HYDROZOA or POLYZOA.** Some naturalists, as above said, have referred the Graptolites to *Bryozoa* or *Polyzoa.* More believe them Hydroid Zoophytes of the Sertularian type.	
G i	*G. sagittarius,* p. 6.	**Graptolithus Hisingeri,** Carruthers (Geol. Mag. 1868, Vol. v.). *G. sagittarius,* Hisinger (Salter, Quart. Journ. Geol. Soc. Vol. VIII. pl. 21, fig. 8). The Skiddaw specimens are only doubtfully referred to this very common Lower Bala species.	a. 33, Haykin Gill. a. 35, Knockmurton in Skiddaw slate. a. 34, Scaw Gill. a. 36, Craig ddu Allt, over Tremadoc Iron Works.
G i	Synopsis, p. 7.	**Diplograpsus mucronatus,** Hall (Pal. N. York, Vol. I. pl. 73, fig. 1). These spinose species of Graptolites are now undergoing much revision, and it is probable great changes will be made in their names. *D. barbatulus,* Salter, is one of them, found at Ty Obry.	a. 37, Ty Obry, east side of Tremadoc Estuary. (A. Sedgwick.)
G i	Synopsis, p. 8.	**Diplograpsus pristis,** His. (Leth. Suec. pl. 35, fig. 5). A most common species, with only slightly prominent square-ended cells.	Tyddyn Dicwm, Tremadoc Estuary. (A. Sedgwick, 1847.)
G i		**Phyllograptus angustifolius,** Hall (Grapt. Quebec Group pl. XVI. figs. 17—21). A remarkable leaf-like form, four rows of cells being placed crosswise on the stem or axis, instead of two, as in Diplograpsus. (Quart. Journ. Geol. Soc. Vol. XIX. p. 137, fig. 7). *P. typus,* Hall (l. c. pl. 15), is a larger species, but very like this.	a. 45, Skiddaw Slate. (From Bryce Wright.)

Case and Column of Drawers.	Reference to McCoy's Synopsis: and Figures of Genera.	Names and References; Observations, &c.	Numbers and Localities.
G i	Under *Grapt. latus,* p. 4. (Not the figure or description, which belongs to *G. priodon,* Pl. 1 B, fig. 7.)	**Didymograpsus latus,** Salter. The specimens to which I restrict the name are from Skiddaw slate localities, and are broken portions of a twin graptolite. The specific name may stand, as unoccupied.	Scaw Gill, Whiteless, a. 39, Knockmurton. Omit the Builth Bridge locality, which contains only the *Graptolites priodon,* and is Wenlock shale.
G i		**Didymograpsus,** sp. very like *D. latus,* and possibly identical.	Upper Arenig, Whitesand Bay. (Dr. Hicks.)
G i	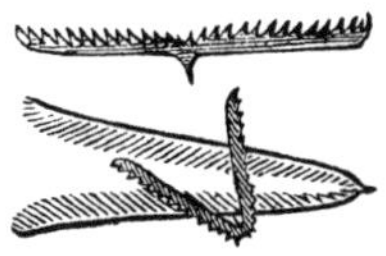	**Didymograpsus geminus,** Hisinger (Leth. Suec. Supp. 2, pl. 38, fig. 3). It is a Stiper Stones species, also found at Whitesand Bay. Mem. Geol. Surv. Vol. III. pl. 11 B, fig. 8.	a. 38, Skiddaw Slate. (From Bryce Wright.)
		Didymograpsus V-fractus, Salter (Quart. Journ. Geol. Soc. XIX. p. 137, fig. 13 e). A species closely related to *D. Pantoni,* McCoy, from Australia.	Skiddaw Slate.
		Didymograpsus Hirundo, Salter (Quart. Journ. Geol. Soc. XIX. p. 137, fig. 13 f). *D. constrictus,* Hall (Grapt. Quebec, pl. 1, fig. 23—27).	Skiddaw, Stiper Stones, Whitesand Bay, St. David's.
G i	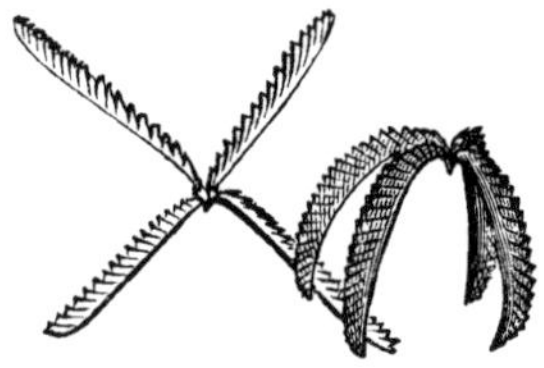	**Tetragrapsus bryonoides,** Hall (*Graptolithus,* ib. pl. III. IV. VI.). A Graptolite with 4 thick branches recurved. Quart. Journ. Geol. Soc. ib. fig. 8 a. The genus is like *Didymograpsus,* but twice branched.	Skiddaw Slate. (From Bryce Wright.)
G i		**Tetragrapsus quadribrachiatus,** sp. Hall (l. c. pl. 5). *Tetr. crucialis,* Salter (Quart. Journ. Geol. Soc. ib. fig. 8 b). Quite a distinct species, with much longer and patent branches.	a. 43, Skiddaw Slate. (From Bryce Wright.)
G i	Compound Graptolites, dichotomously branched on one plane.	**Dichograpsus,** Salter. A large branching Graptolite, first discovered in Canada by Sir W. E. Logan. It is the most compound of Graptolites, excepting *Dendrograpsus* and *Dictyonema.* This is a more compound form of Graptolite than *Tetragrapsus,* the branches bifurcating again and again, but only in one plane. Moreover a horny disk connects the base of the branches in some specimens.	

Case and Column of Drawers.	Reference to McCoy's Synopsis: and Figures of Genera.	Names and References; Observations, &c.	Numbers and Localities.
	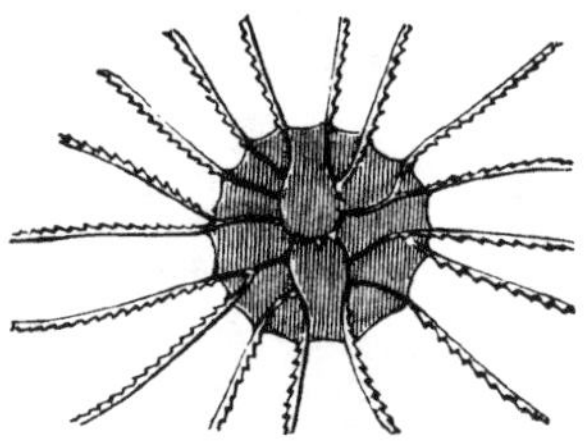	**Dichograpsus octobrachiatus,** Hall (Grapt. Quebec Group, pl. 7, 8). *D. aranea*, Salter	Skiddaw.
G i		**Dichograpsus,** sp. A fine large branching species, with long nodes to the branches.	a. 46, Skiddaw Slate. (From Bryce Wright.)
G i		**Dichograpsus Sedgwickii,** Salter (Geol. Mag. Vol. IV. p. 74. Quart. Journ. Geol. Soc. Vol. XIX. p. 137, fig. 11.)	[Skiddaw Slate. The sp. is easily obtained, and is characteristic.]
G i		**Dendrograpsus furcatula,** sp. Salter (Mem. Geol. Surv. Vol. III. p. 331, pl. 11 a, fig. 5). For genus see again Lower Ludlow Rock. The branches are tufted irregularly and branch repeatedly.	a. 44, Ty Obry, Portmadoc.
G i		**Dendrograpsus arbuscula,** Salter MSS. A small species.	a. 295, Whitesand Bay, St. David's. (Dr. Hicks.)
		ANNELIDA.	
G i	*Worm-burrows.* Pl. 1 A, fig. 1—3.	**Palæochorda,** McCoy. Supposed by some to be marine plants; but evidently the filled-up burrows of marine worms.	Blakefell, Cumberland.
G i		**Helminthites and Scolites,** Salter. The filled-up burrows and surface-trails of marine worms, without impressions of any lateral appendages.	Scaw Gill, Cumberland.
G i	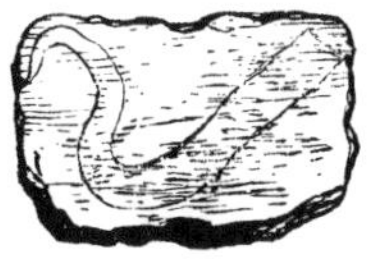	**Scolecoderma** (a worm-tube). See Mem. Geol. Surv. Vol. III. p. 292.	Blakefell.
		CRUSTACEA, PHYLLOPODA AND TRILOBITA.	
G i	*Phyllopods.* 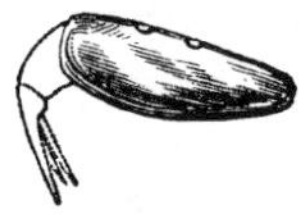	**Caryocaris Wrightii,** Salter (Quart. Journ. Geol. Soc. Vol. XIX. p. 139, fig. 15). A small shrimp-like creature, an inch in length; the carapace 2-valved.	a. 47, Skiddaw Slate. It occurs at Causey Pike and Grassmoor, Cumberland.
		**Primitia?** sp. A bivalve entomostracan. (See Middle Bala for genus.)	a. 48, Ty Obry, Portmadoc. (Mr. Ash.)

Case and Column of Drawers.	Reference to McCoy's Synopsis: and Figures of Genera.	Names and References; Observations, &c.	Numbers and Localities.
G i	*Trilobites.*	**Agnostus Hirundo,** Salter MSS. A large species, with lobed tail-axis.	a. 298, Upper Arenig rocks, Whitesand Bay, St. David's. (Dr. Hicks.)
		Agnostus Morei, Salter (Quart. Journ. Geol. Soc. Vol. XXII. p. 486, woodcut).	Ellergell, near Milburn. [Prof. Harkness' cabinet.]
		Phacops Nicholsoni, id. ib. p. 486.	Whiteside, near Braithwaite, Keswick. do.
G i		**Calymene vexata,** Salter (with *Asaphus Menapiæ*). Undescribed yet. A small species. *Calymene* is very common in Arenig rocks.	a. 469, Ramsey Isle, N. end. (Dr. Hicks.)
G case		**Calymene ultima,** Salter (*C. Ramseiæ,* Hicks). Undescribed.	do. do.
G i		**Calymene Tristani,** Brongniart (Quart. Journ. Geol. Soc. Vol. XX. p. 286), &c. A common French species.	Carn Goran, S. Cornwall.
G case	Synopsis, Pl. 1 F, fig. 7, p. 167.	**Calymene parvifrons,** Salter. One of the very earliest species known of this common Cambrian genus. It has the smallest glabella known in the group. (Mem. Geol. Surv. Vol. III. p. 325. Salter, Mon. Brit. Tril. 1865, pl. 9, figs. 25—28).	a. 56, Tai-hirion, W. of Arenig. (Sedgwick and Salter, 1843.)
G i		**Dionide atra,** Salter (Mem. Geol. Surv. Vol. III. p. 321. pl. 11 A, fig. 9). Allied to *Ampyx* and *Trinucleus,* but with no perforated fringe.	b. 360, Ty Obry, Portmadoc. (Mr. Homfray.)
G i		**Trinucleus Gibbsii,** Salter (Siluria, 2nd ed. p. 53, Foss. 53. fig. 7). The genus is distinguished by its perforated fringe round the glabella and cheeks.	a. 52, Whitesand Bay, St. David's. (J. W. Salter.)
G i		**Ampyx Salteri,** Hicks MSS. Much like *A. nudus,* Murchison.	a. 303, Upper Arenig, Whitesand Bay, St. David's. (Dr. Hicks.)
G h. 10	Pl. 1 G, fig. 24.	**Homalonotus,** sp. (*H. bisulcatus,* Mem. Geol. Surv. Vol. III. p. 328. pl. 11 A, fig. 8). A genus more prevalent in the Upper Cambrian. Like *Calymene* it has 13 body segments, and a forked labrum.	a. 57, Ty Obry. (Mr. Ash.) b. 355, specimen figured in Mem. Geol. Surv. (Mr. Homfray.)
G i		**Homalonotus monstrator,** Salter. Undescribed as yet. *Homalonotus,* like *Calymene,* ranges into Devonian rocks.	a. 482, Ramsey Isle. N. end. Quite the base of the Group. (Dr. Hicks.)

Case and Column of Drawers.	Reference to McCoy's Synopsis: and Figures of Genera.	Names and References; Observations, &c.	Numbers and Localitiea.
G case		**Æglina caliginosa,** ib. (Mem. Geol. Surv. Vol. III. p. 318, pl. 11 A, fig. 10). The genus is known by its large eyes covering the cheeks.	Ty Obry, Portmadoc (abundant), Mr. Ash. b. 350, 351. (Mr. Homfray.)
G i		**Æglina Boia,** Hicks. A small smooth species, the head and tail like *Agnostus.*	a. 297, Whitesand Bay, St. David's. (Dr. Hicks.)
G case	*Æglina.*	**Æglina binodosa,** Salter (Mem. Geol. Surv. Vol. III. p. 317, pl. 11 B, fig. 3. Decade XI. Pl. 4, fig. 1—6). Two tubercles on the third body segment.	W. of Stiper Stones, Shelve. (Mr. Lightbody.)
G i		**Æglina grandis,** id. (Mem. Geol. Surv. Vol. III. p. 317, pl. 12, fig. 11). Two inches long; with rounded tail and granulated surface to head.	a. 53, Whitesand Bay, St. David's. (Dr. Hicks.)
G i		**Ogygia Selwynii,** id. (Mem. Geol. Surv. Vol. III. p. 313, pl. 11 B, fig. 5). The most characteristic trilobite of the Arenig group; and the first one found by Prof. Sedgwick in 1843.	a. 54, W. of the Stiper Stones. (Mr Lightbody.) a. 55, Tai-hirion, W. of Arenig. (A. Sedgwick and J. W. Salter.)
G i	*Ogygia.*	**Ogygia peltata,** id. (Mon. Brit. Tril. 1865, pl. 25*, fig. 2). A large fine species, resembling much the great trilobites of this genus in the slates of Angers. (*O. Desmaresti,* &c.)	a. 50, Whitesand Bay, St. David's. (Specimen figured in Mon. Brit. Tril.)
G i		**Ogygia** ———, sp. long narrow axis to tail.	a. 356, Whitesand Bay. (Mr. Homfray.)
G i case		**Ogygia bullina,** id. (Mon. Brit. Tril. ib. fig. 1). The glabella is much swelled in front and narrowed behind.	a. 49, Whitesand Bay. (Specimen figured in Mon. Brit. Tril.)
G i		**Asaphus solvensis,** Hicks MSS. A small species with smooth tail-piece.	a. 293, Tremenheere, near Solva; Lower Arenig rocks.
G i		**Asaphus Menapiæ,** Hicks (undescribed). A large species, with smooth tail-piece. *Asaphus* becomes abundant in the Arenig group.	a. 469, 470, Ramsey I., St. David's, N. End. Base of group. (Dr. Hicks.)
G i	*Asaphus.*	**Asaphus,** sp. with ribbed tail.	Upper Arenig rocks, Whitesand Bay. (Dr. Hicks.)
		MOLLUSCA.	
G i		**Strophomena,** sp. fine striæ.	Carn Goran, Cornwall.
G i		**Lingula petalon,** Hicks MSS.	a. 294, Arenig rocks of Whitesand Bay.

Case and Column of Drawers.	Reference to McCoy's Synopsis: and Figures of Genera.	Names and References; Observations, &c.	Numbers and Localities.
G i	*Brachiopod Shells.*	**Orthis Carausii,** Salter (Davidson, Earliest Brach. Geol. Magazine, Vol. v. pl. 16, fig. 23). Short, squarish, with 16 simple ribs.	a. 485, Ramsey I., N. End. Base of the whole group. (Dr. Hicks.)
G i		**Orthis** ———, sp. (*O. calligramma* of some works, but a distinct species).	Carn Goran, Cornwall, in hard quartzite.
G i		**Orthis Menapiæ,** Hicks (Davidson, l. c. pl. 16, figs. 24—28). The fine ribs distinguish this from the species above described.	a. 291, Lower Arenig rocks, St. David's. (Dr. Hicks, 1869.)
G case		**Obolella? plumbea,** Salter (*Lingula*, Mem. Geol. Surv. Vol. III. pl. 11 B, fig. 10: Davidson, Sil. Brach. pl. 4, figs. 20—27). *Obolella* is a genus only ranging from the lower part of the Middle Cambrian to the base of the Upper Cambrian.	a. 57, White Grit Mine, Shelve. (Mr. R. Lightbody.)
G i		**Obolella plicata,** Hicks (Davidson, l. c. pl. 4, f. 6).	a. 292, Tremenheere, St. David's. (Dr. Hicks.)
G i		**Obolella,** sp. small and round. The species of all the horny Brachiopods are very difficult to separate one from the other.	b. 359, Ty Obry, near Garth, Portmadoc. (Mr. Homfray.)
G i	*Lamellibranchs.*	**Palæarca socialis,** Salter (Mem. Geol. Surv. pl. 11 A, fig. 13, p. 344). The genus is better represented in the Bala rocks (see Middle Bala).	b. 359, Ty Obry, Portmadoc. (Mr. Homfray.)
G i case		**Ctenodonta elongata,** Hicks (undescribed). The genus like *Nucula*, but the ligament external.	a. 469, Ramsey I., North End. (Dr. Hicks.)
G i case		**Ctenodonta rotunda,** Hicks (undescribed).	a. 469, Ramsey I., N. End. (Same slab as last.)
G i		**Theca Harknessi,** Hicks (undescribed). The surface finely reticulated by striæ in two directions.	a. 299, Upper Arenig rocks, Whitesand Bay. (Dr. Hicks.)
G i	*Pteropods.*	**Conularia Corium,** Salter (Mem. Geol. Surv. Vol. III. p. 355, pl. 11 A, fig. 11). A very large and thin species, with perfectly smooth sides, and a squarish section. Like *C. pyramidata*, Desl. from Normandy, and possibly the same.	b. 363, Ty Obry, Garth. (Mr. Homfray.)
G i		**Orthoceras,** small sp. The genus is rare in this formation.	b. 349, Ty Obry, do. (Mr. Homfray.)

Upper Cambrian (Cambro-Silurian of some authors; Lower Silurian of Murchison; &c.).

As there has been much controversy respecting the name to be borne by the rocks overlying the primordial or "Ffestiniog" group, and underlying the "May Hill Sandstone;" a table of equivalents is here added, which may serve to guide students to the cases, and harmonize the modern text-books with the 'Synopsis' published in 1851-2 by Prof. Sedgwick and McCoy, which is the basis of this Catalogue.

Sir R. I. Murchison. 1831—1859.	Sir H. de la Beche, and Geol. Surv. 1840—1866.	Prof. Phillips. 1855.	Barrande, to 1860.	Sir W. Logan and E. Billings, to 1866.	J. W. Salter. 1864—1866-9.	Sedgwick. 1832—1853.
		Wenlock.			Upper Wenlock. Lower Wenlock.	Wenlock Group. Denbigh flag.
Llandovery rocks, in 1859.	Upper Caradoc, or Upper Llandovery (Upper Silurian).	May Hill.	Etage E. base.	Clinton Group.	May Hill Group.	May Hill Sandstone.
				Anticosti Group.	——	——
Caradoc Sandstone and Llandeilo flags (Lower Silurian).	Lower Llandovery. Caradoc or Bala.	Llandovery. Caradoc or Bala.	Etage D. 5.	Hudson River Gr. Trenton Group.	Llandovery Group. Caradoc or Bala Group.	Upper Bala.
			Etage D. 4.			Lower Bala.
	Llandeilo.	Llandeilo.	Etage D. 3. Etage D. 2.	Black River Group. Bird's Eye Group.	Llandeilo Group.	
	Lower Llandeilo.		Etage D. 1.	Quebec Group. Chazy Group.	Arenig or Skiddaw Group.	Arenig Group.
Lingula flags (base of Lower Silurian).	Tremadoc.	Tremadoc.	Blank here, not noticed by M. Barrande.	Potsdam Group.	Upper } Lower } Tremadoc.	Tremadoc Group.
	Lingula Beds.	Ffestiniog Group.	Etage C.	St John's Group.	Upper } Middle } Ffestiniog Group. Menevian Group.	Ffestiniog Group.
Cambrian, or Longmynd rocks.	Cambrian.	Harlech and Bangor Group.	Etage B.	Huronian.	Harlech Group. Longmynd Group.	Harlech Group. Longmynd Group.

Upper Cambrian (Lower Bala Group: Middle Bala: Upper Bala).

The 'Lower Bala' of the Synopsis (Introd. p. xx.) was made to include the dark earthy slates, with occasional bands of limestone, such as are exhibited on the east flank of the Arenig range; the Mynydd Tarw and Craig-y-glyn above Llanarmon, in the Berwyn Mountains, and the black slates about Bangor and the flanks of Snowdon. It also comprehended the *arenaceous* deposits on the W. side of Bala lake, *below* the Bala limestone, and that limestone itself.

Further research, however, by the Geological Survey has shewn that these dark earthy slates, with occasional limestone, are the equivalent of the black slates and limestones of S. Wales, collectively known as the *Llandeilo flag* in its modern signification*. Prof. Sedgwick therefore permits me to restrict the term to the

* The original 'Llandeilo flag' comprehended much more, and many higher beds, of different ages; and could not, at the time of the publication of the 'Synopsis,' be at all accurately identified. It is much better understood now.

earthy slates and limestones which contain, in favourable localities, a distinct fauna. They are directly comparable with the Orthoceratite limestones of Sweden, the Bird's Eye and Black River limestones of N. America. The Moffat group of S. Scotland and the *upper* Skiddaw slate are part of the series.

The new term, *'Middle Bala' group,* is adopted here for the Bala limestone and its associated sandstones and slates, several thousand feet thick in N. Wales, but reduced to a minimum in S. Wales, where it appears as dark incoherent schist. In Shropshire this series is known as the Caradoc sandstone, with its Horderly limestone. The 'Oskarskal' group of Sweden (Regio VI. *Trinucleorum* of Prof. Angelin, the 'Stratum quartum' of Linnæus) represents this group. It is the Trenton and Hudson River group of N. America, and the major part of Barrande's great fossil-bearing formation Etage D. belongs to it. (The Coniston limestone: the limestone of Kildare: the Craig Head limestone of Ayrshire: the Peebles limestone: all are of this age.)

'Upper Bala' comprehends the Aber Hirnant beds above the Bala limestone, with a peculiar set of fossils: the lower portion of the Coniston flag, viz., that conformable to the limestone; and indeed all beds above the Bala limestone and beneath the May Hill sandstone. In the Bala and Coniston sections, we do not indeed quite reach the horizon of the Llandovery rocks, with their peculiar fauna—*Petraia, Atrypa, Pentamerus,* &c. But the group 'Upper Bala' was made to include all the beds, whether near Meifod, or Welchpool, or near Llanwddyn, Montgomeryshire, which lie *above* the Bala limestone, and *under* the 'unconformable' cover of the Denbighshire grit and flag.

There is therefore both propriety and symmetry in retaining the name 'Upper Bala' for those beds (having on the whole a distinct fauna) which lie above the Bala limestone. The group includes, in ascending order—

3*c*. UPPER BALA (Sedgwick).

1. *Hirnant* limestone and slate (Sedgw.). = Coniston flag, the lower part only (Ash Gill; Coldwell, &c.) above the Coniston limestone. Not the Brathay or Horton flags.
2. *Llandovery beds:*— Phillips, Salter, &c. (Lower Llandovery, Murchison) Medina Sandstone, North America. So called by Murchison from the locality where they are best exhibited. The group has received much illustration of late years. It is the 'Mathyrafal limestone' near Meifod, of Sedgwick. It skirts all the lower border of the Denbigh flags and grits, from a point a few miles S. of Bala to Builth: and then rising out from under the May Hill group at Llandovery, ranges to the sea in Pembrokeshire.

It is the great fossiliferous group at Haverfordwest. Its parallel in Westmorland has not yet been found*, unless part of the Coniston flag (lower) belongs to it. But in S. Scotland it is conspicuous at Dalquorhan and Mullock in Ayrshire. In Galway the fossiliferous rocks of Maume and Cong belong in great part to it. [But the Irish collection is kept separate, and will be catalogued to follow the British one.]

N.B. In the list of localities in the Synopsis, p. 326, &c., the terms Upper and Lower Bala are sometimes vaguely used, owing to the absence of data at that time for a clear definition of the fossil horizons. It is requested, therefore, that the student will consider all the Bala groups of the Synopsis as one, and consult the collection for their division into the modern groups.

For the placing of many of the fossils under these special geological subdivisions, I am alone responsible. But having studied N. Wales under Prof. Sedgwick, S. Wales under Sir H. de la Beche, and having been engaged for seventeen years working at the Silurian and Cambrian fossils of the Geological Survey, Prof. Sedgwick trusts me to arrange them according to the present state of our knowledge, 1867. The Bala group or Upper Cambrian of Sedgwick therefore consists of—

UPPER CAMBRIAN. (Sedgw.)
- Lower Bala = Llandeilo flag (*Upper Llandeilo,* Geol. Survey, the Arenig being the lower).
- Middle Bala = Caradoc sandstone, and Bala rocks (Geol. Survey and Sir R. I. Murchison).
- Upper Bala = Caradoc shales, Hirnant limestone, and Lower Llandovery rock (Geol. Survey).

All these are unconformably overlaid by the 'May Hill Sandstone' or Clinton group, which forms the base of the Silurian (Sedgwick) or Upper Silurian (Murchison).

* Mr. T. McK. Hughes has lately found the equivalents of the Llandovery rock near Coniston, &c. It consists of mudstone, conglomerate, and beds of *Graptolites* like those of Barrande's Etage E. 1, which surely, for many reasons, is a Llandovery group. J. W. S.

LOWER BALA GROUP (earthy dark slates and limestones). N. and S. Wales. S. Scotland. S. Ireland.

Case and Column of Drawers.	Reference to McCOY'S Synopsis: and Figures of Genera.	Names and References; Observations, &c.	Numbers and Localities.
		Zoophyta or Bryozoa.	
		Graptolithus, Auct. The simple one-sided *Graptolites*, with close-set cells, are the only species now called *Graptolithus*. The *Graptolithus* of Linnæus, Syst. Nat. ed. 1, seems to be our *Diplograpsus* (Carruthers).	
Gk	Syn. Pl. 1 B, f. 2, p. 6.	**Graptolithus Sedgwickii,** Portlock, Geol. Rep. pl. 19, fig. 1—3; Harkness, Quart. Journ. Geol. Soc. Vol. VII. p. 60. A well-marked species, with curious straight spine-like processes to the cell mouths.	Moffat, Dumfries. [A note at p. 366 of the Synopsis explains that the specimens labelled Lockerby are from various localities near Moffat and Beattock Bridge.]
Gk		**Graptolithus Hisingeri,** Carruthers (Geol. Mag. 1868, p. 126). *G. sagittarius*, His. not Linn. There seems little doubt that Linnæus described under this name a fossil plant instead of a graptolite. The species is excluded from all other strata except the Lower Bala (or Llandeilo flag), and is common therein.	Moffat, Dumfries.
G (case).	Pl. 1 B, f. 3, p. 4. (*G. Millepeda*, Pl. 1 B, f. 6, p. 5.)	**Graptolithus lobiferus,** McCoy (*G. Becki*, Barrande). A form with the cell-ends so tumid as to give a very peculiar appearance. (*G. millepeda*, McCoy, is the young thereof, as suggested by himself.)	a. 59, Moffat. a. 58 (*G. millepeda*); do.
	Pl. 1 B, f. 4, 5, p. 6.	**Graptolithus Nilssoni,** Barrande, Grapt. Bohéme, pl. 2, figs. 16, 17. Narrow stem, and short cells.	Moffat, and common in N. Wales.
Gk		**Graptolithus tenuis,** Portlock, Geol. Rep. Londondy. and Tyrone, p. 319, pl. 19, fig. 7. A narrow thin-stemmed short-celled species.	a. 68, Moffat, Dumfries (common in S. Wales also. J. W. S.).
G (case).	(*Grapt. Convolutus*, p. 3.)	**Rastrites (Graptolithus) convolutus,** Hisinger, sp. (Leth. Suec. t. 35, fig. 7). *Rastrites triangulatus*, Salter and Harkness (Quart. Journ. Geol. Soc. Vol. VII. p. 59).	Moffat (not Lockerby), Sedgw. a. 60 (as *G. convolutus*), Moffat, id.
		Rastrites peregrinus, Barr. may be the young of *R. convolutus* (Grapt. Bohéme, pl. 4, fig. 6).	Moffat.

Case and Column of Drawers.	Reference to McCoy's Synopsis: and Figures of Genera.	Names and References; Observations, &c.	Numbers and Localities.
		Rastrites maximus, Carruthers (Geol. Mag. 1868, p. 126). A species with very large cells.	[Moffat.]
	p. 7.	**Diplograpsus Folium,** Hisinger (Leth. Suec. t. 35, fig. 8). The double-ranked Graptolites were first distinguished by McCoy in 1851 as *Diplograpsus*.	a. 63, Moffat, Dumfriesshire. (Sedgw.)
		Diplograpsus angustifolius, Hall (Pal. N. Y. III. p. 515). The young states of Diplograpsus have been shewn by Prof. Hall to be single trispinous cells.	[Moffat (Carruthers).]
Gh or case above.	p. 7.	**Diplograpsus mucronatus,** Hall (Pal. N.Y. Vol. I. p. 268, t. 73, fig. 1).	a. 66, Cairn Ryan, Wigtonshire.
		Diplograpsus pristis, Hisinger? (Leth. Suec. p. 114, t. 35, fig. 5. *Fucoides dentatus,* Brong. *Grapt. foliaceus,* Beck. Sil. Syst.). One of the numerous varieties or species referred to this almost cosmopolitan form. Mr. Carruthers figures it with long thread-like radicles, and believes this to be the true species of Hisinger.	a. 62, Conway Castle, N. Wales. Pen-y-goylan, Llandeilo. Pen Cerrig, Builth.
	p. 8.	**Diplograpsus pristis** *β*, McCoy. A variety figured by Prof. Hall; with broader and more triangular teeth.	a. 72, Moffat.
		Diplograpsus persculptus, Salter MSS. (Carruthers, Geol. Mag. 1868, p. 130). A beautiful undescribed form, very distinct in its sculpture from *D. pristis.*	a. 64, Pumpsant, near Dolaucothi, in Caermarthenshire. (J. W. Salter.)
		Climacograpsus, Hall, cells in double row, but excavated, not projecting. The stem cylindrical (Grapt. Quebec, p. 111).	
	Diplograpsus rectangularis, Pl. 1 B, f. 8.	**Climacograpsus scalaris,** Linn. sp. (*Grapt. scalaris* of Linn. in Skanska Resa, p. 147. (Carruthers, l.c. fully cleared up.) *Grapt. teretiusculus,* Hisinger, Leth. Suec. t. 38, fig. 4, suppt. *Diplograpsus teretiusculus,* Salter, Quart. Journ. Geol. Soc. Vol. VIII. pl. 21, figs. 3, 4. [*D. rectangularis,* McCoy, is only a variety with closer cells].	Anglesea. Moffat. a. 61 (*D. rectangularis*), Moffat, McCoy.

Case and Column of Drawers.	Reference to McCoy's Synopsis: and Figures of Genera.	Names and References; Observations, &c.	Numbers and Localities.
		Climacograpsus bicornis, Hall (Grapt. Quebec, p. 111).	[Extremely common in the Lower Bala.]
Gk	p. 9.	**Didymograpsus sextans.** A common American species. The twin branches (δίδυμος) in this genus of Prof. McCoy's diverge at once from a minute radicle: and have the cells on the inner edges of each branch: but in this species and many others the stems diverge so widely as to be bent backwards (Pal. N. Y. Vol. I. p. 273, t. 74, fig. 3).	a. 65 (*Diplograpsus?* McCoy), Cairn Ryan, Wigtonshire.
G and Gk	Not of p. 5, nor in the Synopsis.	**Didymograpsus Murchisoni,** Bœck. Siluria, 2nd ed. pl. 1, fig. 1. An extremely common Lower Bala (Llandeilo flag) species in N. and S. Wales—especially at Abereiddy Bay. The branches are strictly parallel and are broad.	a. 64, Abereiddy Bay, Pembrokeshire. (J. W. Salter.)
G and Gk	Pl. 1 B, f. 11, 12, p. 10.	**Protovirgularia dichotoma,** McCoy. A very interesting form, branched like *Dichograpsus,* but with large cells or groups of cells in double rows; and according to Prof. McCoy, very like the living *Virgularia.*	a. 71, Near Moffat, Dumfriesshire. (Prof. Sedgwick.)
	Corals.	**ZOOPHYTA** proper. (Actinozoa, Actinaria, Zoantharia, Cœlenterata of various authors).	
G and Gk	p. 24.	**Stenopora fibrosa,** Goldf. sp. (Pet. Germ. pl. 28, figs. 3, 4). The commonest of all Millepore corals in the older Palæozoic rocks.	Golden Grove, Llandeilo,
Gk		**Stenopora ramulosa,** Phill. sp. (Mem. Geol. Surv. Vol. II. pt. 1, p. 385 as var. 1 of *S. fibrosa.*)	Tregib, Llandeilo.
Gk		**Nebulipora favulosa,** Phill. (Mem. Geol. Surv. Vol. III. p. 282, pl. 19, fig. 10). A more irregular species than the *N. lens,* so common in Middle Bala rocks. The coral consists of minute tubes, of which the fertile ones are in enlarged clusters (*Monticulipora,* Edw. & Haime, is a nearly contemporary name).	Llandeilo.
Gk	*Palæopora interstincta,* p. 15.	**Heliolites interstincta** Linn. sp. (Edw. & Haime, Pal. Monogr. pl. 57, fig. 9). It is doubtful if this be the species, but so named by McCoy.	a. 69, Llandeilo.

Case and Column of Drawers.	Reference to McCoy's Synopsis: and Figures of Genera.	Names and References; Observations, &c.	Numbers and Localities.
		ECHINODERMATA.	
G case and Gk		Crinoid, and probably Cystidean fragments, but of unknown species and genera, are common in the limestone of Craig-y-Glyn and Craig-y-Beri, Llanrhaiadr.	a. 70, Craig-y-Glyn, Llanrhaiadr.
	Worm tubes and tracks.	**ANNELIDA.**	
G case Gk	Pl. 1 B, f. 13, p. 10.	**Pyritonema fasciculus,** McCoy. The analogy between this bundle of tubes and the skeleton of the glass plant polype is very close. But the real affinity is most probably with the straight-tubed gregarious *Serpulæ,* such as are found in the carboniferous rocks, *Serpula parallela* for instance. In proof of this there occurs another species, which has the tubes separate.	a. 73, Tregib, Llandeilo. [This is the specimen on which was founded the supposed *fish-defence* from the Llandeilo flag. The resemblance is very great at first, but I had no excuse for the blunder, which misled others. J. W. S.]
Gk	*Serpulites dispar,* p. 132 in part only.	**Pyritonema** ———, sp. Loose tubes in greenish volcanic grit.	Tan-y-craig, N. of Builth. a. 76. Also in Shale, at Tre-coed, 3 m. N. of Builth.
		N.B. The following 4 species are in all lists included in the Lower Bala or Llandeilo flag, but it is quite probable they are of Middle Bala age.	
Gk	p. 128.	**Nemertites Olivantii,** McLeay (Sil. Syst. pl. 27, fig. 4). Superficial trails of long worms (*Helminthites?*).	a. 72, Llampeter, S. Wales.
G case	p. 129.	**Myrianites McLeayii,** Murchison (Sil. Syst. pl. 27, fig. 3). Impression of the trail of a rather long worm, in zigzag folds. At one end we seem to have the impression of the animal itself, with its lateral feet or bundles of setæ.	a. 75, Llampeter.
G case Gk	p. 129.	**Nereites Sedgwickii,** McLeay. Sil. Syst. pl. 27, fig. 2. The large foliaceous gill-plates and cirrhi are strongly shewn in this imprint. It is the original figured specimen.	a. 74, Llampeter. [N.B. This species also occurs at Aberystwith, in Middle Bala rocks.]

Case and Column of Drawers.	Reference to McCoy's Synopsis: and Figures of Genera.	Names and References; Observations, &c.	Numbers and Localities.
G case	*Worms* continued. p. 129.	**Nereites Cambrensis,** McLeay, (Sil. Syst. t. 27, fig. 1). It is a mistake to represent this as a very long worm. Its trails are visible for some length: four or five individuals on the slab having traversed some space (marked by a simple line) and been imbedded at the end of their trail. In one case the worm has again retreated, before death, along the line made by his track.	a. 72*, Llampeter, S. Wales, the figured specimen. Probably Middle Bala. The rolling system of S. Wales consists, as we now know, of Upper and Middle Bala rocks.
		Crustacea. Phyllopods and Trilobites:— No higher orders than bivalved and Apus-like *Entomostraca,* with Trilobites, have been detected in any beds beneath the May Hill Sandstone.	
Gk G case	*Bivalve Crustacea.* Pl. 1 L, f. 2 (*Cytheropsis*).	**Cythere** (*Cytheropsis*) **Aldensis,** McCoy. A minute smooth bivalve shell, slightly curved. Such species are abundant everywhere; and have been mostly left for description to Prof. R. Jones.	a. 77, Aldeans Limestone. Stinchar River. a. 77*, Lower Caradoc Shales, Shineton, Buildwas.
		Primitia, one or two species (Prof. R. Jones).	Lower Shales, Shineton, near Buildwas, Shropshire.
Gk	*Beyrichia,* Pl. 1 E, f. 3, p. 136.	**Beyrichia complicata,** Salter, Mem. Geol. Surv. Vol. II. pt. 1, pl. 8. fig. 16, p. 295, Vol. III. pl. 19, fig. 9. *Beyrichiæ* are bivalved Crustacea, with lobed and furrowed carapaces.	Pont-y-meibion, Llanarmon, N. Wales (coll. by Prof. Sedgwick and J. W. Salter, 1842).
	Trilobites.	**Trilobita.**	
Gk	*Diplorrhina triplicata.* Syn. Pl. 1 E, f. 11, p. 142.	**Agnostus MacCoyii,** Salter. Decades Geol. Surv. XI. pl. 1, figs. 6, 7. Mem. Geol. Surv. Vol. III. p. 297, pl. 13, fig. 8. Very common in Llandeilo flag. Only generally, not minutely like the Bohemian species figured by Barrande and Corda. *Agnostus* is the simplest of all trilobites.	a. 86, Pen Cerrig, N. of Builth. It occurs everywhere, but sparingly in S. Wales, in these rocks.
Gk		**Agnostus** ———, sp. not quite perfect enough to determine.	Shineton, Buildwas.
Gk	(*T. gibbifrons* and *T. Caractaci* in part, pp. 144, 5.)	**Trinucleus favus,** Salter (Mem. Geol. Surv. Vol. III. p. 320, pl. 13, fig. 9). This is distinguished by the square form and honeycomb-structure of the fringe, the outer cells being largest;—like those of drone-cells in a comb.	a. 980, Llandeilo. Pres. by Mr. T. McK. Hughes. From Tregib, and Golden Grove, Llandeilo. (McCoy has labelled some of these *T. gibbifrons*). Craig-y-beri, Llanarmon fach.

Case and Column of Drawers.	Reference to McCoy's Synopsis: and Figures of Genera.	Names and References; Observations, &c.	Numbers and Localities.
G	*Tretaspis,* p. 146. Pl. 1 E, f. 16. As *var.* of *T. gibbifrons,* p. 145.	**Trinucleus fimbriatus,** Murchison (Siluria, 2nd ed. pl. 4, fig. 7). A species with a narrow furrowed border; the puncta in rows. *Trinucleus* is a simple trilobite: no eyes or facial suture. The seemingly complex punctate border is remarkable, but very easily explainable as a series of spines connected by transverse processes.	3 m. North of Builth (as *T. gibbifrons,* var.)
G G	p. 148. Pl. 1 E, f. 13, p. 147.	**Ampyx nudus,** Murch. Decades Geol. Surv. 2, pl. 10. A remarkable genus like *Trinucleus,* but with long front- and side-spines to the head. *Ampyx latus,* McCoy, is only a var. of the preceding, distorted.	a. 101, Builth. [The Tregib locality is erroneous.] *A. latus,* Builth. Abereiddy Bay, Pembrokeshire. (Dr. Hicks.)
G k	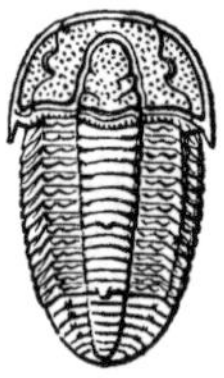	**Conocoryphe monile,** Salter, n. sp. Glabella lobed, the front marginal furrow dotted.	b. 167, Shineton Shales, Buildwas.
G k	*Conocoryphe.*	**Conocoryphe** or **Triarthrus.** A small species, probably of *Triarthrus,* an Upper Bala genus in N. America. This was formerly thought an *Olenus;* and the shales were therefore called Olenus shales.	Shineton Shales, Buildwas. [It is possible that the locality may include some Tremadoc beds.]
G k		**Æglina,** sp. The genus has six body segments, and enormous eyes covering the side of the head. It is European only, and strictly a Bala fossil (Lower and Middle Bala).	do.
	Æglina.	**Æglina,** sp. imperfect. The species are numerous.	do.
G	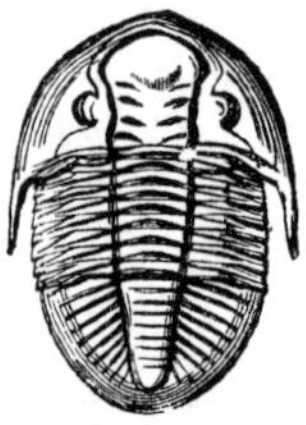	**Ogygia corndensis,** Murchison (Salter, Mon. Brit. Tril. p. 130, pl. 16). A broad axis for the genus.	a. 84, Builth.
G	p. 148. *Ogygia.*	**Ogygia Buchii,** Brongniart (Salter, Mon. Brit. Tril. pl. 14, 15). Male and female specimens of this, the flattest of all Trilobites, and probably with the most immoveable segments.	Wellfield, Builth; and Pen Cerrig, Builth.
G k		**Ogygia angustissima,** Salter, Mon. Brit. Tril. p. 129, pl. 14, figs. 8, 9, may be only a variety of *Ogygia Buchii.*	[Builth. Presented by Mr. T. McK. Hughes.]

Case and Column of Drawers.	Reference to McCoy's Synopsis: and Figures of Genera.	Names and References; Observations, &c.	Numbers and Localities.
G	*Ogygia*, p. 149. Pl. 1 F, f. 2.	**Barrandia** (Ogygia) **radians.** McCoy (Salter, Mon. Brit. Tril. 1866, pl. 19, figs. 1—4). A genus distinguished from *Ogygia* by the length of the glabella, and its shape.	a. 80, Pen Cerrig, Builth. (A most prolific locality, full of the fry of Trilobites.)
G	Synopsis, Pl. 1 F, f. 1, p. 149.	**Barrandia Cordai,** McCoy (Salter, Mon. Brit. Tril. 1866, pl. 19, fig. 5). Short, oval, smooth.	a. 79, Pen Cerrig, Builth.
G k		**Asaphus?** sp. with a short tail-piece.	Shineton, Buildwas.
		[**Lichas patriarchus,** Edgell, Geol. Mag. 1866, p. 162. The earliest species of this genus known in Britain. See Middle Bala for genus.]	Pont Ladies Quarry, Llandeilo.
G	Pl. 1 F, f. 14, p. 155.	**Cheirurus (Eccoptochile) Sedgwickii,** McCoy. A remarkable, large, trilobite, more expanded than any other of the British forms; they are mostly very convex. A kindred species occurs in Bohemia.	a. 78, N. of Builth.
G	*C. brevicapitata*, Pl. 1 F, f. 4. *C. Baylei*, Pl. 1 F, f. 8, p. 165. Pl. 1 G, f. 25, 28.	**Calymene Cambrensis,** Salter (Mon. Brit. Tril. pl. 9, figs. 12—14). This has been included under *C. Baylei*, Barrande, and also *C. brevicapitata* Portl. by McCoy; but the last is a Middle Bala species, and no Bohemian forms are identical with ours.	a. 97, Llandeilo.
G k	Pl. 1 E, f. 18 *a*, p. 170.	**Homalonotus bisulcatus,** Salter (Mon. Brit. Tril. 1865, pl. 10, figs. 3—10). *Homalonotus* was a long-lived genus, and of all the more perfect trilobites the least trilobed.	Porth Treuddyn, Caernarvonshire (in beds supposed to be = the dark earthy slates of the Ffestiniog slate quarries).
G k		**Asaphus** (Isotelus) **laticostatus,** McCoy, not of Green. The *Asaphus laticostatus* of Green is a *Phacops*. (See Salter, Mon. Brit. Tril. 1866, p. 158, pl. 18, fig. 6.)	a. 208, Maen Goran, Builth.
G h	p. 171.	**Asaphus tyrannus,** Murch. (Salter, Mon. Brit. Tril. pl. 21, 22.) One of the largest of all known trilobites: sometimes a foot or more in length. Mr. Hughes' specimen (981) must have been rather more. It is as characteristic of the Lower Bala as the *A. Powisii* is of Middle Bala rocks.	a. 83, Llandeilo. a. 85, Tregib. a. 82, Craig-y-Glyn, Llanarmon, N. Wales. a. 981, Llandeilo. A fine large specimen. (Mr. T. McK. Hughes.)

5

Case and Column of Drawers.	Reference to McCoy's Synopsis: and Figures of Genera.	Names and References; Observations, &c.	Numbers and Localities.
G	*A. tyrannus, labrum.*	**Asaphus peltastes,** Salter (Mon. Brit. Tril. pl. 22, figs. 1—4). A smaller and more convex species, with larger eyes, a narrower head, and fewer tail segments.	a. 81, Llandeilo. Abundant with the more common large species just mentioned.
G	Also *I. latus* (part), p. 172 (not the figure).	**Illænus crassicauda,** Wahl.? (*Ill. latus* McCoy in part). A caudal shield only: and probably the only specimen known—of this common Swedish species—in British rocks. (See Salter, Mon. Brit. Tril. 1866, p. 181, woodcut, fig. 44.)	a. 90, Bugon, Knockdollian, Ayrshire. (As *I. latus*), same locality.
		Mollusca Brachiopoda.	
G k	*Lamp shells.*	**Discina?** sp. Species of this genus occur everywhere in all rocks, ancient and modern.	Pen Cerrig, Builth.
G k	*Pseudo-Crania*, p. 187. Pl. 1 H, f. 1.	**Crania divaricata,** McCoy. More common in Middle Bala rocks: the free unattached species of *Crania* may deserve separation.	Tan-y-craig, Builth.
		Lingula Ramsayi, Salter (Dav. Sil. Brach. p. 55, pl. 3, figs. 49—52). A large pentagonal species.	Abereiddy Bay (Dr. Hicks).
G	p. 252.	**Lingula granulata,** Phillips, Mem. Geol. Surv. Vol. II. pt. 1, p. 370, pl. 25. A fine square species, with granular lines of growth, crossed by regular fine rays.	Llandeilo, rare. Meadowtown, Shelve.
G	p. 253.	**Lingula obtusa,** Conrad? (Hall, Pal. N. York, t. 30, fig. 7, Davidson, Monog. Sil. Brach. 1866, p. 52, pl. 3, figs. 31, 32.) Doubtful if it be the American species—or a form of the following—	a. 104, Llandeilo.
G		**Lingula attenuata,** Sowerby. (Siluria, 2nd ed. pl. 5, fig. 16. Dav. Sil. Br. pl. 3, figs. 18—27.)	a. 2, Llandeilo.
G	p. 251.	**Lingula curta,** Conrad? Pal. N. York, t. 30, fig. 6. Considered to be a doubtful species.	a. 103, Wellfield, Builth.
		[**Spondylobolus? craniolaris,** McCoy. See Wenlock rocks; it is clearly not a Lower Bala fossil,—but from the Wenlock shales.]	

Case and Column of Drawers.	Reference to McCoy's Synopsis: and Figures of Genera.	Names and References; Observations, &c.	Numbers and Localities.
G k	Pl. 1 H, f. 3, p. 188. 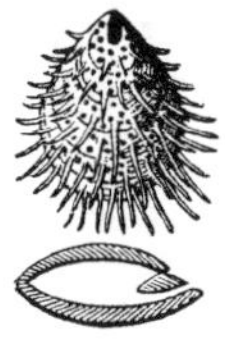	**Siphonotreta micula,** McCoy (Davidson, Mon. Sil. Brach. pl. 8, figs. 2—6). A pretty small Brachiopod, covered with spines, and very nearly allied to the *Discinæ.* Larger species, with thicker shells, occur in the Lower Bala rocks of Russia and Sweden. Also in the Upper Silurian of Britain.	a. 102, Pen Cerrig, and Wellfield, near Builth, Radnorshire.
G k	*Spirifera,* p. 192.	**Orthis biforata,** Schlotheim: a variety with fine ribs. This common species occurs over all the northern hemisphere.	a. 100, Tregib, Llandeilo. Rare in Lower Bala.
G k	*O. parva,* p. 221.	**Orthis elegantula,** var. *parva,* Pander. A variety only of the commonest of all British species of *Orthis.* It is comparatively rare in Lower Bala.	Nant-yr-Arian, Llandeilo.
	p. 214.	[**Orthis striatula,** Emmons (Mem. Geol. Surv. Vol. III. pl. 13, figs. 10—14)].	Common in Lower Bala.
G k	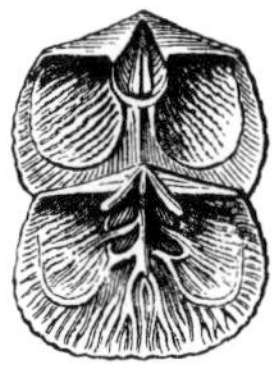	**Orthis calligramma,** Dalm. (Mem. Geol. Surv. Vol. III. pl. 22, fig. 1). This common shell ranges from Russia to Britain; and from the Arenig to the Wenlock rocks.	Tan-y-graig, Builth (in volcanic grit). a. 98, Cerrig Cregyn, Anglesea (quoted p. 215, as Upper Bala).
G k	Pl. 1 H, f. 20—24, p. 229.	**Orthis turgida,** McCoy. A species with a remarkably tumid upper valve.	Golden Grove, Llandeilo, in grit (also quoted as Upper Bala). Llandeilo.
G case G k	p. 215. 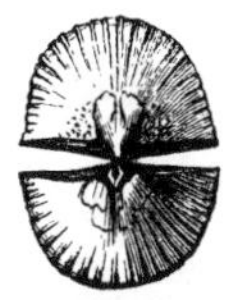	**Orthis confinis,** Salter (Quart. Jour. Geol. Soc. v. t. 1, fig. 4, Davidson, l. c. pl. 36, figs. 1—4). A large transverse species, with fine ribs. Is *O. sagittifera* the same species ?	a. 99, Craig-y-beri, Llanarmon, N. Wales. a. 96, Bugon, Knockdollian, Ayrshire. Llandeilo.
	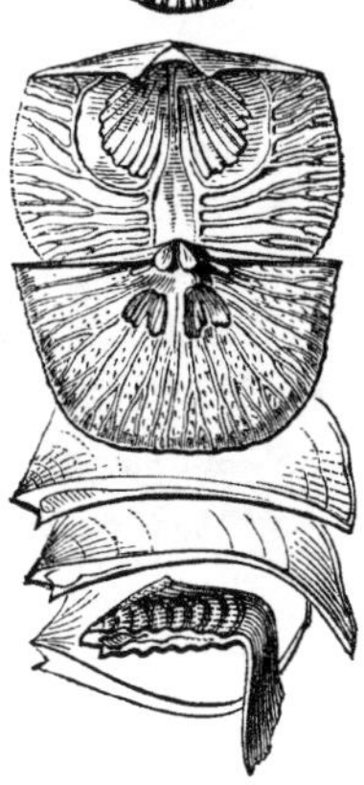	**Strophomena.** Rafinesque, Davidson, Salter. Differs from *Orthis* by having a widely expanded thin shell, and the cardinal process double.	
G k		**Strophomena concentrica,** Portlock, Geol. Rep. p. 452. A shell extremely common in the Caradoc or Middle Bala rocks.	Craig-y-beri, and Y Foel fawr, N. of Llanrhaidr.
G k		**Strophomena,** sp. (S. expansa, McCoy in part). A very flat species, with a much thinner shell than Sowerby's *Stroph.* (*Orthis*) *expansa.* The latter is a Middle Bala sp.—this is a common Lower Bala one.	a. 105, Nant-yr-Arian, Llandeilo.

Case and Column of Drawers.	Reference to McCoy's Synopsis: and Figures of Genera.	Names and References; Observations, &c.	Numbers and Localities.
G k		**Strophomena,** sp. with distant radiating ribs.	Bugon, Knockdollian, Ayrshire.
	L. tenuissimestriata.	**Leptæna.** Dalman, Salter, Davidson, in a stricter sense than used by McCoy. Valves involute, muscular impressions greatly elongated. See Wenlock List, postea.	
G k	Pl. 1 H, f. 44, p. 239. p. 237.	**Leptæna sericea,** Sowerby. Siluria, 2nd ed. pl. 5, fig. 14. The commonest of all shells in British Upper Cambrian, apparently rare out of Britain.	a. 94, Llandeilo.
G k		**Leptæna,** sp. A pretty species with wavy ornaments between the ridges.	a. 92, Colmonel, on the Stinchar, Ayrshire.
	Palæarca.	**Lamellibranchiata** (ordinary bivalves) are not uncommon in sundry portions of the Llandeilo flags—but none are yet in the collection. They abound much more in Middle and Upper Bala rocks, in all countries.	
	Orthis biforata.	The student may be reminded of the essential difference between the Lamellibranchiate and Brachiopod shells, by keeping in mind that the shells of the former are placed laterally, and are unequal on each side of the umbo, in the latter they are dorsal and ventral, and the umbo is truly central. The dorsal valve is really the analogue of the foot of the Lamellibranch. (Salter.)	
G		**Mollusca Gasteropoda.**	
	p. 292.	Univalves, of a few types only, are common in Lower Bala rocks.	
G		**Murchisonia** (*Hormotoma*) **angustata,** Hall, Pal. N. York, t. 10, fig. 2. The beaded whorled species (*Hormotoma,* Salter), are common in Bala rocks, especially in America.	a. 93, Bugon, Knockdollian, Ayrshire.
G k		**Pleurotomaria,** sp.	Same locality.
G k	*Trochus helicites,* p. 297.	**Raphistoma æqualis,** Salter. Siluria, 2nd ed. woodcut foss. 37, fig. 2 (*R. qualteriata,* Mem. Geol. Surv. Vol. III. p. 271). A common genus in the Bala group—flat discoid shells.	a. 95, Llandeilo.

Case and Column of Drawers.	Reference to McCoy's Synopsis: and Figures of Genera.	Names and References; Observations, &c.	Numbers and Localities.
G k		**Euomphalus,** sp.	S. W. of Pwllheli (Mid. Bala ?)
	Pl. 1 L, f. 15, p. 301.	**Eccyliomphalus? scoticus,** McCoy. The true Eccyliomphalus of Portlock is a Pteropod; this may be *Phanerotinus* of Sowerby.	a. 91, Knockdollian, Ayrshire.
		Heteropoda. Blainville (*Nucleobranchiata*, Rang). All or most living forms of this group are thin light shells; but there seems much reason for Dr. Woodward's opinion, that the heavy *Bellerophon* and *Maclurea* were solid representatives of this low order of Molluscs.	
G	Pl. 1 L, f. 13, p. 300, as *M. magna.* 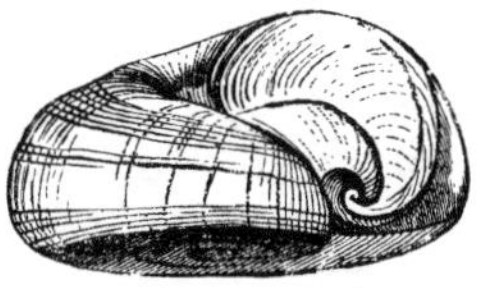 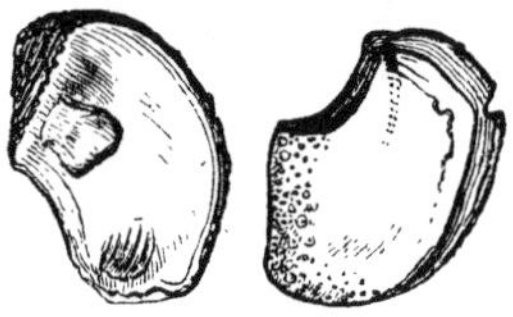 *Operculum of Maclurea.*	**Maclurea Logani,** Salter (Geol. Survey, Canada Decade I. pl. 1). The proportions of the mouth and whorls of this massive shell agree better with *M. Logani* than the great *M. magna* of the Chazy limestone.	a. 89, Colmonel, Stinchar River, Ayrshire. It also occurs at Bugon, and other Ayrshire localities.
		Cephalopoda, Cuvier. The tetrabranchiate families (only) are known in Palæozoic rocks—and Nautiloid and Orthoceras-like forms are among the very earliest known Molluscs, except Pteropoda and Brachiopoda. They are particularly abundant in the representatives of the Llandeilo flag or Lower Bala groups.	
G	p. 312. 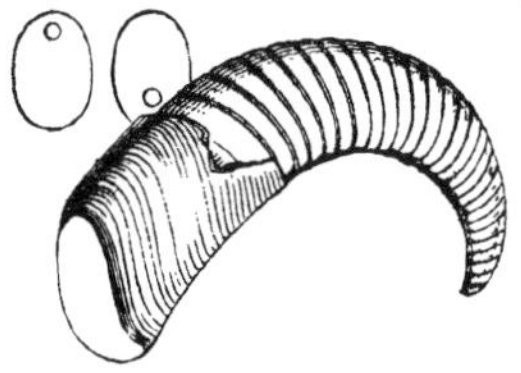 *Cyrtoceras.*	**Cyrtoceras multicameratum,** Hall. Hall's sp. is from Trenton limestone, which represents Middle Bala. Many forms migrated in Cambrian times eastward from America, and are consequently of older date there than in Britain. But few follow a reverse order of progression.	a. 88, Knockdollian Mt., 3 miles from Ballintrae, Ayrshire.
G	p. 314. 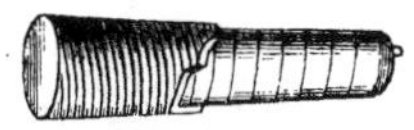	**Orthoceras** (*centrale*, Hisinger ?), Leth. Suec. t. 9, fig. 4. Not likely to be Hisinger's Swedish species.	a. 87, Llandeilo.
G k 5	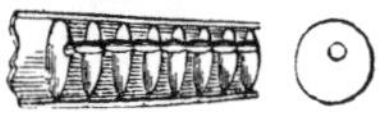 *O. subundulatum,* p. 317.	**Orthoceras fluctuatum,** Salter, n. sp.——Coarser striæ than *O. subundulatum,* Portl. and apparently more bent still than in that species.	a. 611, Wellfield, Builth, in the hard volcanic flag.

Lower Bala. Irish Collection.

Dark earthy slates (with Graptolites) of Wexford, and parts of Waterford and Clare. The lower Bala is not greatly fossiliferous in Ireland, and the Lingula flags and Tremadoc are absent, so far as yet known; the Lower Bala resting on the Harlech Group in Wicklow.

Case and Column of Drawers.	Reference to McCoy's Synopsis: and Figures of Genera.	Names and References; Observations, &c.	Numbers and Localities.
	Graptolites. p. 7.	**Diplograpsus mucronatus,** Hall (Pal. N. Y. Vol. I. p. 268). Whether this species be truly named seems matter of doubt. The cells are closer than in the specimens from Wigtonshire.	Belvoir, Clare. Gibbet Hill, Tinnaglough, Wexford. (Geol. Surv. Ireland.)
	p. 8.	**Diplograpsus ramosus,** and perhaps **D.** *pristis.*	Waterford (Ballintray). (Geol. Surv. Ireland.)
	Trilobites.	**Barrandia Portlocki,** Salter. *Ogygia Portlocki* in Decades Geol. Surv. No. 2, pl. 7. *Asaphus dilatatus,* Portlock, Geol. Rep. t. 24, figs. 1, 2, 7 (not the rest); *Barrandia* in Salter's Monograph, Brit. Tril. pl. 19, figs. 6—10. The short tail axis and broad glabella distinguish *Barrandia.*	Newtown Head, Waterford. Both Lower and Middle Bala appear to occur in this important locality. (Sir R. Griffiths.)
		Orthis striatula, Emmons (Mem. Geol. Surv. Vol. III. pl. 13, figs. 10—14). The flat form, and fine striæ curved up to the hinge line, mark this as distinct from *O. testudinaria,* its companion, and the teeth are different from those of *O. elegantula.*	Newtown Head, Waterford.

MIDDLE BALA GROUP. Arenaceous rocks and slates (volcanic grits and schaalsteins, with beds of felstone porphyry (Snowdon, Moel Hebog, &c.).

Extent of the group. Prof. Sedgwick only includes in this group the 9000 feet of beds, chiefly arenaceous, slaty, and with some calcareous bands, which lie over the dark earthy slates of the Arenig section (Lower Bala). The group extends a short distance, probably a couple of hundred feet, above the Bala limestone. But it does not include the Hirnant limestone, which is the base of the Upper Bala group next described. It appears to represent in mass the whole of the Caradoc Sandstone proper, in Shropshire.

Organisms. The Hudson River group of New York—the Utica Slate and Trenton limestones—are parallel to this large British group. In Sweden, Region D. of Angelin. In Bohemia, Etage D. 3, 4, 5. This group is the richest in organic life of all the groups beneath the Old Red Sandstone, not excepting the Wenlock. Most of the orders of invertebrata have been found in it; but the supposed remains of fish have proved mythical. No *Eurypteridæ* among the *Crustacea*, have yet been found. But Trilobites reach their maximum here, both as to genera and species (see Mon. Brit. Tril. 1864, Introduction, p. 8). Brachiopods are various and innumerable. Corals, Cystideæ, and Crinoids are also plentiful, though not of many species, except as regards the *Cystideæ.* These attain their maximum here in Britain. Sponges; cup and millepore corals; starfish of several species; tubicolar and naked Annelida; Entomostraca—are all common. Mollusca of all orders (except naked Cephalopods)—the Heteropods and Pteropods being of giant size. The bivalve shells (Lamellibranchs) are allied to Modiola, Arca, and Avicula: (see Phillips, Mem. Geol. Surv. Vol. II. Pt. 1, p. 264).

Case and Column of Drawers.	Reference to McCoy's Synopsis: and Figures of Genera.	Names and References; Observations, &c.	Numbers and Localities.
		Zoophyta? or Bryozoa. GRAPTOLITIDÆ. I think these are Bryozoa (Mem. Geol. Surv. Vol. III. Appendix, p. 328), but arrange them in the order of the Synopsis, as their systematic position is somewhat doubtful.	
Gk7	*G. Ludensis.* p. 4.	**Graptolites priodon,** Bronn (*G. ludensis,* Murch. Siluria, 2nd ed. p. 64, t. 12, fig. 1). The finest specimens of this long-lived fossil are from rocks of this group in Scotland.	Grieston, Peebleshire. Penarth uchaf, N.W. of Pen-y-glog.
Gk7	p. 5.	**Graptolites sagittarius,** Hisinger, (not Linn.) Leth. Suec. t. 35, fig. 6. Salter, Siluria, 2nd ed. p. 542, Quart. Journ. Geol. Soc. VIII. p. 390. (*G. Hisingeri,* Carr. Geol. Mag. V. p. 126.)	Horton, Ribblesdale; Rother Bridge, Studgill.
Gk7	p. 6, Pl. 1 B, f. 2.	**Graptolites Sedgwicki,** Portl. (Siluria, 2nd ed. Foss. 11, fig. 2). A common species in the Tyrone and Fermanagh schists of Ireland, which are most probably of this age.	Grieston, on the Tweed.
Gk7		**Graptolites,** sp. with very narrow cells.	Bala. (J. Peters, Esq.)
Gk7	p. 6, Pl. 1 B, f. 4, 5.	**Graptolites tenuis,** Portl. Geol. Rep. p. 319, t. 22, f. 6. (Siluria, 2nd ed. p. 55, Foss. 10, fig. 12.)	Grieston, on the Tweed.

Case and Column of Drawers.	Reference to McCoy's Synopsis: and Figures of Genera.	Names and References; Observations, &c.	Numbers and Localities.
G k 7	p. 8.	**Diplograpsus pristis,** Hisinger, sp. Leth. Suec. t. 35, f. 5. Siluria, 2nd ed. Foss. 11, fig. 4.	a. 210, Ardwell, Girvan.
G k 7	p. 8.	**Diplograpsus var. β,** McCoy.	Hollies, Church Stretton? (more probably Soudley, J. W. S.). Horderley.
	Sponges.	**Amorphozoa.**	
G k 7		**Astylospongia,** sp. (*grata,* Salter, MSS.). One of the lobed sponges. The genus is one of Dr. Ferd. Rœmer's (Geol. of Texas).	a. 211, Coniston (on a Ctenodonta). Pwllheli, N. Wales.
G k 7	p. 12.	**Stromatopora striatella,** D'Orb. Siluria, 2nd ed. p. 240, fig. 51. A sponge? formed of calcareous spiculæ arranged at right angles to the outer layer which is formed by the fusion of their apices.	a. 212, Coniston.
G k 7		**Sphærospongia hospitalis,** Salter (for genus see Strachey's Geology of India), for sp. see Catalogue Mus. Pract. Geology.	Onny River. (R. Lightbody, Esq.)
G k 7		**Ischadites?** minute tesselated surface—*micropora,* n. sp.	a. 213, Blaen-y-cwm, Llansantffraid, Glyn Ceiriog, N. Wales.
G k 7		**Ischadites?** sp. wavy ridges.	a. 214, do.
G k 7		**Vioa (Cliona),** sp. filling up borings by an annelide in the shell of a Brachiopod.	a. 215, Horderly.
		Zoophyta—Anthozoa. *Zoantharia Tabulata.*	
G k 7		**Heliolites, Dana.** A tabulate (millepore) coral related to the *Heliopora cærulea* of the Australian reefs.	
	Palæopora. Syn. p. 15, Pl. 1 C, f. 3.	**Heliolites favosa,** McCoy sp. The most closely approximated cells in all the genus except a new one in Wenlock list (see postea).	a. 106, Craig Head, Colmonel, on the Stinchar river, Ayrshire.
G k 7	p. 15, 17.	**Heliolites interstincta,** Wahl. The common *Heliolites* of the Wenlock rocks is equally common here.	Coniston Water. Sunny-Brow. Long Sleddale. Aplethwaite Common.
G k 7		**Heliolites interstincta** (as *P. petalliformis* on tablet).	Sunny Brow. Coniston.

Case and Column of Drawers.	Reference to McCoy's Synopsis: and Figures of Genera.	Names and References; Observations, &c.	Localities and Numbers.
	Pl. 1 c, fig. 2. p. 16.	**Heliolites subtubulata,** McCoy. Tubes more distant and the crenulations or septa nearly obsolete.	Sunny Brow. Coniston. Llansantfraid, Glyn Ceiriog.
Gk bottom drawer.	p. 18.	**Heliolites tubulata,** Lonsd. sp. (Siluria 2nd ed. pl. 39, fig. 3). The prominent edges of the small close cups easily distinguish this.	Coniston.
		Heliolites, sp.	a. 126, Barking, Dent.
Gk	Pl. 1 c, fig. 4. p. 16.	**Heliolites megastoma,** McCoy (Siluria, 2nd ed. Foss. 27, fig. 7. M. Edw. and Haime, Mon. Brit. Fos. Cor. t. 58, fig. 2. Very large cells evenly scattered, mark this fine species—equally abundant in the Dudley limestone.	a. 122, Coniston. High Haume. Maes Meillion, Bala. Blaen-y-cwm, Nantyr, in Glyn Ceiriog.
k	Pl. 1 c, fig. 9, p. 20.	**Favosites crassa,** McCoy. Subcylindrical curved branches like *F. cristata* of the Wenlock.	Coniston.
Gk	p. 19.	**Favosites alveolaris,** Goldf. Pet. Germ. t. 26, fig. 1. Much like the common *F. Gothlandica,* and with variations like that species. Prof. McCoy and myself seem to be the only English palæontologists who recognize this excellent species by its ragged crenulated edges to the tubes.	Dry Ridge, Horton (as *F. crassa*) Llansantfraid, Glyn Ceiriog, Alt goch, Llanfyllyn. Mynydd Fron Frys, W. of Chirk. Cefn-y-coed, Glyn Ceiriog.
Gk	Synopsis, p. 20.	**Favosites Gothlandica,** Linn. sp. Siluria 2nd ed. Foss. 17, figs. 2, 3. A species of coral which has the widest range—from America to Eastern Europe: and from the Lower Bala to the Carboniferous.	Common enough in Bala localities, and easily obtained.
		(Nebulipora) Monticulipora. See Wenlock limestone.	
Gk	Pl. 1 c, fig. 6, p. 23.	**Nebulipora (Monticulipora) explanata,** McCoy.	Coniston.
Gk		**Nebulipora favulosa,** Phillips (Mem. Geol. Surv. III. pl. 19, fig. 10).	Applethwaite Common. Coniston.
Gk	Pl. 1 c, fig. 5, p. 24.	**Nebulipora papillata,** McCoy. (See Wenlock Limestone).	a. 120, Coniston.
Gk	Pl. 1 c, fig. 7, p. 23.	**Nebulipora lens,** McCoy. A hemispheric coral often rising into a subpyramidal form, and covered with unequal clusters of pores—the fertile ones largest—like the drone-cells in the hive.	a. 217, Horderley, Glyn Diffwys. Bala, Cwm of the Cymmerig, Bala; Moel Uchlas, Llanfyllin.

Case and Column of Drawers.	Reference to McCoy's Synopsis: and Figures of Genera.	Names and References; Observations, &c.	Localities and Numbers.
Gk	p. 24.	**Stenopora fibrosa,** Goldfuss, sp. One of the commonest of all corals in the Cambrian rocks. The broad branches present radiating tubes in the fractured sections.	a. 218, a. 121, Coniston. Dry ridge near Horton, St. Helen's. Blaen-y-cwm, Nantyr; Cyrn-y-Brain, Corwen; Alt yr anker, and Penllys, near Meifod; Glyn Diffwys, Bala; Girvan, Ayrshire.
Gk	p. 24.	**Stenopora var. Lycoperdon** (Hall, Pal. N. Y. Vol. I. pl. 23, fig. 1). A hemispherical variety of the above; most common in the Bala rocks; and easily distinguishable from the *Nebulipora lens* above quoted, by the tubes being all of one diameter.	Coniston Water. Moel Uchlas, Llanwyddon; all round Bala; Glyn Ceiriog; Conway; Bwlch-y-groes; Sclattyn Road; Meifod.
Gk	p. 25.	**Stenopora var. β regularis,** McCoy. Lons. sp. Sil. Syst. t. 15, fig. 1.	Blaen-y-cwm, Nantyr. a. 107, Acton Scott, Cader Dinmael.
Gk		**Stenopora var. ramulosa,** Phillips, sp. (Mem. Geol. Surv. Vol. II. Pt. 1, p. 385).	Bryn Eithin, Pemmachno; Bwlch-y-Groes. Llansantfraid, Glyn Ceiriog; Tyn-y-Cabled, Llanfyllin.
		Halysites, Fischer. Common chain coral (see Wenlock).	
Gk	p. 26.	**Halysites catenulatus,** Linn. *Catenipora escharoides* (Goldf. Pet. p. 74, pl. 25, fig. 4). The common chain coral.	a. 124, Coniston; Ingleton and Thornton: Applethwaite. Near Chirk; and Llansantfraid, Glyn Ceiriog in N. Wales.
Gk	p. 26.	**Halysites var. labyrinthica** (Goldf. Pet. pl. 25, fig. 5). Apparently not a distinct species, but a luxuriant variety of the last mentioned.	High Haume; Dalton-in-Furness.
	Cup-Corals.	**Zoantharia rugosa,** Milne Edwards and Haime. This is the first appearance in Britain of the remarkable tribe of corals, a distinct order, with the form of ordinary Turbinoliæ, but the structure of parts characteristic of the *Alcyonidæ*, all the septa are in fours, and the tube divided by transverse tabulæ.	
Gk	*Caninia*, p. 28.	**Omphyma turbinata,** Linn. sp. (M. Edwards and Haime, Mon. Brit. Foss. Cor. p. 287, t. 69, fig. 1). A common coral in the Wenlock rocks, and the longest usually found.	Craig Head, Ayrshire.

Case and Column of Drawers.	Reference to McCoy's Synopsis: and Figures of Genera.	Names and References; Observations, &c.	Localities and Numbers.
G k	*Caninia*, p. 28.	**Omphyma turbinata** var.	Coniston.
G k		**Omphyma turbinata,** (or perhaps *Cyathophyllum*).	Coniston.
G k	(*Strephodes*, p. 30). Pl. 1 c, fig. 10.	**Cyathophyllum Craigense,** McCoy. (The twisting of the lamellæ at the base of the cup is not uncommon in *Cyathophyllum,* and *Strephodes.*)	a. 100, Craig Head, Girvan.
G k	*Sarcinula*, p. 37.	**Sarcinula** (*Syringophyllum*) *organum,* Linn. sp. (Mon. Brit. Fos. Cor. pl. 71, fig. 3). One of the most frequent fossils in Sweden, and Norway, in the slate rocks.	Coniston Water. Long-Sleddale. High Haume.
G k	p. 39. Pl. 1 B, fig. 23, 24. *Petraia.*	**Petraia æquisulcata,** McCoy. The genus Petraia is eminently Silurian and Devonian: and is but rare in Cambrian rocks—except their upper members. Its metropolis is the May Hill Sandstone. Like *Cyathophyllum* in all respects except habit, for the cup is always very deep, and the *tabulæ* so twisted and matted, as to form a solid base. Hence, in casts the solid base disappearing—a cavity is left in the stone: in the cup portion, the matrix assumes the shape of a grooved and striated pyramid, truncated abruptly. The species are numerous, and not yet thoroughly made out. N.B. *P. æquisulcata* has the septa more regular than usual in the genus.	Coniston; Bala.
G k	p. 40.	**Petraia elongata,** Phill. (Pal. foss. Cornwall and Devon. t. 2, fig. 6 B), Bala.	Bala.
G case at end.	p. 40.	**Petraia rugosa,** Phill. (id. t. 2, fig. 7).	a. 204, Penarth, Meifod, on the same tablet with *P. subduplicata* in the Llandovery collection (end case).
G k	Pl. 1 B, fig. 26, p. 40.	**Petraia subduplicata,** McCoy. An extremely frequent fossil in the Upper Bala (Llandovery) rocks and easily distinguished by its regular form and crenulated septa. (*P. ziczac* McCoy, Sil. foss. Ireland, p. 60, appears to be the same.)	Llansantfraid, Glyn Ceiriog.

Case and Column of Drawers.	Reference to McCoy's Synopsis: and Figures of Genera.	Names and References; Observations, &c.	Numbers and Localities.
G case	Pl. 1 B, fig. 25, p. 41.	**Petraia uniserialis,** McCoy, a small short species.	a. 203, Llansantfraid (on tablet with *P. subduplicata,* as above, case G).
	(*Lace-corals, &c.*). Phyllopora.	**Polyzoa (Bryozoa,** Ehrenberg). Placed here, as a high type of the Radiata (Cœlenterata, Huxley) by Prof. McCoy. Naturalists usually, and with good reason, allot them a place with the molluscoids—near the Tunicate group. The Lace coral and most of the deep sea minute corals, as they are called, are really Bryozoa. But they assume so much of the aspect of branching corals, that it is convenient to arrange them with them in the cabinet—especially as in fossil forms it is not always possible to separate these two widely remote groups. A Polyzoon is nearer to a *Terebratula* than to the coral it imitates.	
G 1	*Berenicea.* Pl. 1 C, fig. 17, p. 45.	**Berenicea heterogyra,** McCoy, a flat patch of cells, growing on shells.	Coniston.
G 1		**Berenicea heterogyra,** var.	Cader Dinmael, W. of Corwen.
G 1	p. 46. Pl. 1 C, fig. 15.	**Ptilodictya costellata,** McCoy (Genus *Stictopora* of Hall). The genus consists of branched fronds, flat or rather slightly convex on each side with regular cells on either face. The quincunx arrangement of these gives an elegant pattern in the cast. There are very many species—and the genus ranged over the world in Palæozoic time.	a. 219, Llansantfraid, Glyn Ceiriog. Girvan, Ayrshire.
G 1	Pl. 1 C, fig. 14, p. 47.	**Ptilodictya fucoides,** McCoy. A narrow-leaved form.	Llansantfraid, Glyn Ceiriog, Corwen, Bala, several localities.
G 1	Pl. 1 C, fig. 16, p. 46.	**Ptilodictya explanata,** McCoy. A broad-leaved, crisped and undulated frond, with large cells.	Llansantfraid, Mynydd Fron Frys, near Chirk. Cyrn-y-Brain.
G 1	(*P. acuta,* p. 45 also).	**Ptilodictya dichotoma,** Portlock, Geol. Rep. Londonderry and Tyrone, pl. 21, fig. 3), *St. acuta* of Hall is not the same as ours—but yet occurs in Britain.	a. 117, Coniston. Llansantfraid, Glyn Ceiriog.

Case and Column of Drawers.	Reference to McCoy's Synopsis: and Figures of Genera.	Names and References; Observations, &c.	Localities and Numbers.
G 1		**Ptilodictya acuta,** var. *minor*, Hall. (See Pal. N. York, Vol. I. pl. 26, fig. 3.)	Llechwedd Llwydd, Llansantfraid, Glyn Ceiriog.
	Lace-corals.	**Phyllopora** (King), *Retepora* of most authors. To this genus (which consists of lace-corals with the pores in *many* rows covering the outer surface of the cup-like frond) all the Reteporas of the Palæozoic rocks are referred. They all differ from the living Retepores by having a calcareous layer on which the cells are set.	
G	*Retepora*, p. 48. Pl. 1 C, fig. 18.	**Phyllopora Hisingeri,** McCoy. A very common species.	a. 118, Cefn Coedog, S. of Corwen: Cyrn-y-Brain, Wrexham: Mynydd Fron Frys, Chirk, Coniston.
G 1		**Phyllopora,** sp.	Meifod.
G i		**Phyllopora,** sp.	a. 220, Coniston (High Haume).
G 1	p. 49. *Crinoids.*	**Fenestella Milleri,** Lonsdale. The genus has much the appearance of *Retepora*, but is calcareous, and has cells only on the main rods; not on the cross-bars (fig. see Wenlock).	Llansantfraid, Glyn Ceiriog. Cefn Credog, Corwen.
G 1	p. 50.	**Fenestella subantiqua** (D'Orbigny, McCoy). *F. antiqua* Lonsd. Sil. Syst. pl. 15, fig. 16.	Llansantfraid, Glyn Ceiriog.
		Echinodermata. *Starfish, Crinoids* or Sea-lilies, and *Cystideæ* (Grape-lilies, Globe-crinoids).	
G 1	Pl. 1 D, fig. 4. *Globe-Crinoids, Cystideæ.*	**Glyptocrinus? basalis,** McCoy. (*Cœlocrinus*, Salter, Mem. Geol. Surv. Vol. III. pl. 23, fig. 4.) The characteristic crinoid in Britain of the slate rocks—easily distinguished by its hollow stem.	a. 127, Alt-yr-Anker, Meifod.
		Echinosphærites (Caryocystites) granatum, Wahl.	Rhiwlas.
G 1		**Echinosphærites (Caryocystites) Davisii,** McCoy.	a. 125, 6. Coniston.
G 1	Pl. 1 D, fig. 5, p. 61.	**Sphæronites stelluliferus,** Salter (Mem. Geol. Surv. Vol. III. pl. 20, fig. 6 (*S. aurantium*, Forbes, in part only).	Sholes Hook, Pembrokeshire.

Case and Column of Drawers.	Reference to McCoy's Synopsis: and Figures of Genera.	Names and References; Observations, &c.	Localities and Numbers.
G 1	*Sphæronites.*	**Sphæronites,** sp. The group of the *Cystideæ* is amazingly characteristic of Cambrian rocks—*i.e.* of Upper Cambrian or Bala strata. All the genera with pores scattered over the whole surface belong to this horizon. Those with a web-like ornament, viz. *Echinosphærites,* and those with linked scattered double pores (Sphæronites) are of this age (Lower Silurian of most authors). Those with few large rhombs to contain the pores and limit them belong, on the other hand, to Upper Silurian rocks. None rise higher, and very few traces indeed occur in the Middle Cambrian.	Sholes Hook.
	Protocystites. *Star-fishes.* *Asteriadæ.* *Palæaster.*	**Asteriadæ.** All the starfishes of the Cambrian and Silurian rocks belong to a peculiar division, which differs from ordinary starfish by having the plates which border the avenues of suckers of larger size than the rest. In the Bala species this character is very conspicuous.	
G 1	Quoted by mistake as an (Upper) Silurian species at the foot of p. 60. p. 59.	**Protaster Salteri,** Forbes. (Mem. Geol. Surv. Vol. III. pl. 23, fig. 3.) A species of *Protaster* (or *Tæniaster,* Billings. which is a kindred genus) so very like the Brittle stars (*Ophiura*) of the present day that it was so described by Forbes in Vol. I. Quart. Journ. Geol. Soc. 1845, p. 20. See Mem. Geol. Surv. above quoted, note to p. 290. See also Annals Nat. History, 2nd ser. Vol. 20, pl. 9.	a. 221, Pen-y-gair, Cerrig-y-Druidion. (This unique specimen, collected by Prof. Sedgwick and Mr. Salter in 1844, was lost for 18 years, and then recovered.) The *Protasters* were so named and described by Forbes in Decade I. of the Geol. Survey.
		Palæaster, Hall (*Urasterella,* McCoy, MSS. = Stenaster, Billings.)	
G 1		**Palæaster obtusus,** Forbes (Mem. Geol. Surv. III. pl. 23, fig. 1). A species with very thick blunt arms, and a small mouth.	Bala Lake (foot of). One specimen first described by Forbes, found by Prof. Sedgwick—and then others by the Geol. Survey.

Case and Column of Drawers.	Reference to McCoy's Synopsis: and Figures of Genera.	Names and References; Observations, &c.	Numbers and Localities.
G 1		**Palæaster asperrimus,** Salter (Mem. Geol. Surv. III. pl. 23, fig. 2), with long blunt arms, and a very rough upper surface.	a. 222*, Cast. The original from near Welsh Pool, in Mus. Pract. Geology.
G 1		**Palæaster ?** same species crushed (*Ast. primæva* of Salter (not of Forbes) Decade I. Mem. Geol. Surv.), see list in Quart. Journ. Geol. Soc. Vol. I. p. 20, as before.	a. 222, Moel-y-garnend. W. side of Bala Lake.
G 1		**Palæaster squamatus** (Salter MSS.), upper side only.	a. 110, Bala. (Rev. J. Peters.)
G 1		**Protaster Petri** (Salter MSS.), a new form, very like *P. Sedgwickii,* Forbes.	a. 108, Bala. (Rev. J. Peters.)
G 1	Worm-trails.	**Annelida**—Traces, &c.	
G 1	p. 129.	**Nereites Sedgwickii.** This is the impression of the worm itself, and shews the broad lateral processes (*elytra*). The feet of marine worms are simple processes, with bunches of *setæ* or bristles, and sometimes fleshy *cirrhi.*	a. 223, Thornielee Quarry, Tweed. Aberystwith. (Dr. Milligan.)
case G	p. 129.	**Nereites cambrensis,** McLeay var. a. (Silur. Syst. pl. 27, fig. 1).	Thornielee Quarry, Tweed.
G	p. 130. Pl. 1 D, fig. 13.	**Myrianites tenuis,** McCoy. Here we have only the numerous coiling trails made by a narrow worm-like body in the fine mud.	a. 224, Grieston-on-Tweed, Inverleithen. [N. B. I strongly suspect that the worms *Nereites McLeayii, N. Sedgwickii,* and *Nemertites,* from Llampeter are of this geological age—Middle Bala.]
	Worm-tubes, *horny and shelly.* *T. annulatus,* p. 63.		
G		**Tentaculites anglicus,** Salter (Siluria, 2nd ed. pl. 1, fig. 3). Straight unattached worm tubes, resembling some horny Mediterranean species: and none still like the *Cornulites serpularius* of the Wenlock and Dudley rocks.	a. 128, Horderley. The Hollies, Shropshire. Llansantfraid, Glyn Ceiriog, N. Wales. Ravenstonedale, Westmorland.
G	Pl. 1 D, fig. 10, p. 133. See Upper Ludlow rock.	**Trachyderma ? lævis,** McCoy. (I strongly suspect this to be a fragment of *Serpulites longissimus;* and to have come from Upper Ludlow rock.)	(Said to be from) Acton Scott, Shropshire.

Case and Column of Drawers.	Reference to McCoy's Synopsis: and Figures of Genera.	Names and References; Observations, &c.	Numbers and Localities.
G	*Serp. dispar*, p. 132. Pl. 1 D, figs. 11, 12.	**Serpulites**, sp. (not the *S. dispar*, Salter) from Upper Silurian.	Caradoc grit, Hollies, Shropshire?
G 1	Pl. 1 D, fig. 15, p. 130.	**Crossopodia scotica,** McCoy.	a. 225, Thornielee Quarry, on Tweed.
	Small bivalve Crustacea (water-fleas).	**CRUSTACEA—Phyllopoda and Trilobita.**	
G 1	*Phyllopoda.* Pl. 1 E, fig. 6, p. 138.	**Ceratiocaris? umbonatus,** Salter. Oval convex valves of a bivalve Crustacean, very common.	Bala; Corwen; Conway Falls; Llanfwrog, near Ruthin, under the Old Red Strata (stream section). a. 113, Dermydd Fawr, Denbighshire.
G 1		**Primitia McCoyii** (Salter MSS. — Cythere phaseolus, McCoy, Synopsis Sil. Foss. Ireland, p. 58, pl. 1, fig. 1).	Keisley, Dufton, Westmorland.
G 1		**Primitia McCoyii,** var.	Pusgill, Dufton, ib.
G 1	Pl. 1 E, fig. 3, p. 136.	**Beyrichia complicata,** Salter (Mem. Geol. Surv. Vol. III. pl. 19, fig. 9), a most curiously ornamented and deeply grooved species, extremely common.	Coniston. Pwllheli. Llanfwrog, Ruthin; Dermydd Fawr. Pont-y-Meibion, S. of Llangollen, Mynydd Mawr, Llanfyllin, &c. &c.
G 1		**Beyrichia complicata,** var.	Dermydd Fawr, Denbighshire.
	Pl. 1 E, fig. 1, p. 136.	**Primitia strangulata,** Salter. (One of the simpler forms of the genus—with only one furrow.	Coniston.
G 1	*Trilobites.*	**Trilobita** auctorum. The majority of large forms of this order (always excepting the giant *Paradoxides* of the Lingula flags) occur in the Bala group: great species of Asaphus, Homalonotus, Lichas, &c. In fact the Bala group may be taken to be the metropolis both of the Trilobites, the Orthides, and the Orthoceratites.	
G 1	*Trinodus agnostiformis,* Pl. 1 E, fig. 10, p. 141. *T. tardus,* Pl. 1 F, fig. 9, p. 142.	**Agnostus trinodus,** Salter (Decade Geol. Surv. No. 11, pl. 1, figs. 8—10). The simplest form of trilobite known—and the smallest. It represents well the embryonic forms of larger trilobites.	Rhiwlas, Bala.

Case and Column of Drawers.	Reference to McCoy's Synopsis: and Figures of Genera.	Names and References; Observations, &c.	Numbers and Localities.
		N.B. [*Trinodus tardus* of Barrande, though very like ours has sufficient distinction. None of the Mid-European trilobites are the same as the British.	
Gk		**Trinucleus,** Llhwyd, Murchison. Of all the curious genera of Trilobites, this is one of the most curious, so far as the head portion is concerned; for there is nothing peculiar about the body. A perforated border, which border consists of a double plate, surrounds the whole head, in front and sides: and is really built up of numerous fringing spines connected by their margins into one piece. There are no eyes (except some obscure traces in the subgenus *Tretaspis*), and scarcely a trace of a facial suture. The genus began with the earliest Bala rocks (Upper Cambrian) and must be said to have died out with them—only one or two straggling specimens occurring in higher beds, and this very rarely. On the contrary, every member of the Bala rocks and every locality has more than one species of what might be called the " Lace trilobite." It could roll into a ball, like the woodlouse.	
G	*T. Caractaci*, p. 144, var. *elongatus*, p. 145, *T. radiatus*, p. 146, *T. gibbifrons*, p. 145, Pl. 1 E, fig. 14. *T. latus*, p. 145, Pl. 1 E, fig. 15. *Tretaspis fimbriatus*, p. 146, Pl. 1 E, fig. 16.	**Trinucleus concentricus,** Eaton (Salter, Mem. Geol. Surv. Decade 7, Pl. 7, p. 5). The commonest of all the species from Ohio to Russia, it is liable to much variation, and has received many names. Some of the varieties are really worth notice, such as *T. elongatus*, Portlock. In attempting to separate these forms, the author of the Synopsis has entangled himself in the difficulties attending the study of fossils in cleaved and distorted strata. All those here noticed are mere states of preservation. *T. gibbifrons* represents the ordinary form in Wales, and the uncompressed Caradoc of Shropshire. Some of the specimens have fewer pores than others, &c. &c. But the species is neatly distinguished by the form of the fringe, viz. flat and horizontal above,	a. 133, as *T. Caractaci*, Cheney Longville shales, and Horderley, Shropshire. Llansantfraid; Llechwedd, Llwydd; Bwlch-y-groes; Llanwddyn; Pwllheli, Bala; Llanfechan. As *T. radiatus*, N. of Tremadoc (in beds of Bala age, supposed to be in upper Tremadoc slate !). p. 337. As *T. elongatus* (not of Portlock), Pwllheli. As *T. gibbifrons*, Dolydd Ceiriog waterfall in the Berwyns; Rhiwargor; Garnedd Fawr; W. of Bala; a. 134, Dinas Mowddwy;

Case and Column of Drawers.	Reference to McCoy's Synopsis: and Figures of Genera.	Names and References; Observations, &c.	Numbers and Localities.
		and keeled below, with the pores in quincunx. All the other species differ from this in the arrangement of the fringe.	a. 134*, Bwlch-y-groes; Llanwddyn, Conway River; Glyn Ceiriog. As *T. latus*, Bala; Maes Meillion; Yspytty Evan; Selattyn; Pen Cerrig Serth. As *T. fimbriatus*, Bryn Melyn; Cefn hir fynydd.
G	*Tretaspis*, p. 147. 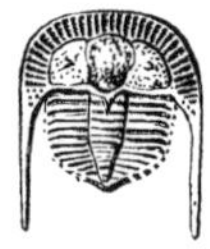	**Trinucleus seticornis,** Hisinger (Leth. Suec. Supp. p. 3, t. 37, figs. 2, 3). See also Decade VII. pl. 7, p. 7, where the surface is described as reticular, and the fringe bent down from the glabella suddenly.	Dufton, Westmorland; Cefn Grugos; Rhiwargor; Rhiwlas.
Gk	*L. propinqua*, p. 150. *L. subproquinqua*, Pl. 1 F, fig. 17.	**Lichas laciniatus,** Dalman. A Swedish species.	a. 138, Coniston.
Gk	p. 337. 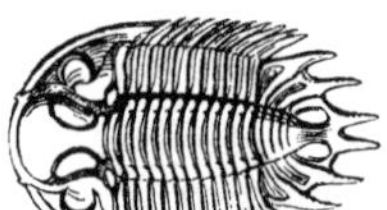	**Lichas laxatus,** McCoy (Synops. Sil. foss. Irel. pl. 4, fig. 9. Salter, Mem. Geol. Surv. Vol. III. pl. 19, figs. 1—3).	Llanwddyn; Cefn Goch; Glyn Ceiriog; Rhiwlas.
Gk	*Trochurus*, p. 151, Pl. 1 F, fig. 16.	**Lichas (Trochurus) nodulosus,** Salter. A very knotty tailpiece, the interstices between the ribs being puffed out into knobs.	b. 231, Pont-y-Glyn, Diffwys, W. of Corwen.
Gk	p. 152. 	**Acidaspis Brightii,** Murch.? (Siluria, 2nd ed. pl. 18, figs. 7, 8). It is not certain that this is the species figured originally from Dudley, and common there.	a. 112, Blaen-y-cwm, Nantyr, S.E. of Corwen. (Beds apparently below the Bala limestone, but possibly above it, J. W. S.)
Gk	Pl. 1 F, fig. 15, p. 153. 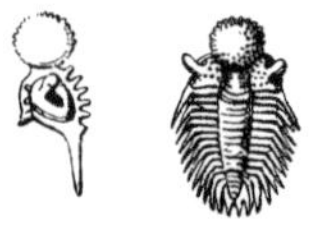	**Staurocephalus Murchisoni,** Barr. (Mem. Geol. Surv. Dec. XI. pl. 5, figs. 1—4). One of the genera related to *Cheirurus*, and with a globular glabella (stomach) placed far in front—with stalked eyes, a body of 11 rings, and a comb-shaped tail piece. Occurs in Wenlock limestone.	b. 230, Rhiwlas, Bala.
	Ceraur. Clavifrons, Pl. 1 F, fig. 12. 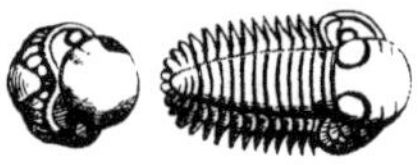	**Sphærexochus boops,** Salter, Mon. Brit. Trilob. p. 79, pl. 6, figs. 27, 28.	a. 114, Applethwaite Common; Coniston.

Case and Column of Drawers.	Reference to McCoy's Synopsis: and Figures of Genera.	Names and References; Observations, &c.	Numbers and Localities.
Gk	*Ceraurus Clavifrons,* Pl. 1 G, fig. 9, p. 154. 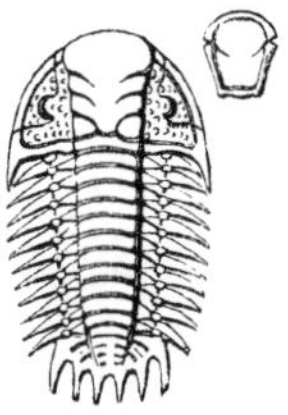	**Cheirurus juvenis,** Salter (Mon. Brit. Tril. p. 67, 1864, pl. 5, figs. 9—12, Mem. Geol. Surv. III. p. 323, pl. 18, figs. 1, 2). *Cheir. clavifrons* of Dalman has been proved distinct, and *Ceraurus* is an exploded genus of Green. N.B. The head or rather glabella is much inflated in this section of the genus, termed *Actinopeltis* by Corda.	b. 193, Bala; Cader Dinmael; Cefn Grugos, near Llanfyllin.
	Ceraurus, Pl. 1 G, figs. 10, 10 *a*.	**Cheirurus octolobatus,** McCoy, sp. (Salter, Mem. Geol. Surv. Vol. III. p. 323, pl. 18, fig. 3. Mon. Brit. Tril. pl. 5, figs. 13, 14). *Ceraurus,* McCoy, Synopsis, p. 154.	b. 229, Rhiwlas, N. of Bala.
Gk	*Zethus atractopyge,* Pl. 1 G, fig. 1—5, p. 156.	**Cybele verrucosa,** Dalm. (Mem. Geol. Surv. 1866, Vol. III. p. 324, pl. 19, fig. 7). *Cybele* is an excellent genus established by Lovèn. *Zethus* is quite a different form, *Z. verrucosus* of Pander and Volborth is only a Russian species. Ours is quite common in Britain and Sweden.	a. 115, Coniston; Ravenstonedale; Horton; Alt-yr-Anker, Meifod. (It was one of the earliest trilobites figured by Dr Llhwyd. 1698).
Gk	*Zethus,* Pl. 1 G, fig. 8. 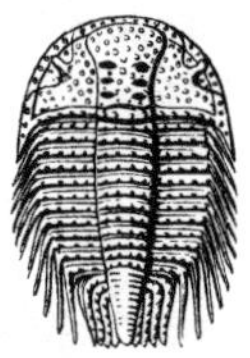	**Cybele rugosa,** Portlock (first figured as Ogygia! by Portlock; then described as Cybele by Salter in Mem. Geol. Surv. Vol. II. Pt. 1. See Morris, Catal. 1854, p. 103). The spines of the tail stream backward so much as to reach far beyond the axis.	a. 116, Coniston (J. Marshall, Esq.).
k	*Zethus,* Pl. 1 G, fig. 6, p. 157. 	**Encrinurus sexcostatus,** Salter (Mem. Geol. Surv. Vol. III. p. 324, pl. 19, figs. 5, 6, Decade VII. pl. 4, figs. 1—11). *Encrinurus* has no produced spines to the ribs of the thorax: and generally it is a more compact form than Cybele, yet closely allied. *Zethus* is quite distinct from both, and if admitted at all (it was founded on a *Cheirurus*) can only be applied as above.	Bala limestone, Rhiwlas.
Gk	*Encrinurus.*	**Encrinurus multiplicatus,** Salter (with many side ribs to tail).	a. 226, Barking, Dent, W. Yorkshire.
Gk		**Phacops,** Emmrich. The most perfect and typical of the whole trilobite family or order, with a compact form in the majority	

Case and Column of Drawers.	Reference to McCoy's Synopsis: and Figures of Genera.	Names and References; Observations, &c.	Numbers and Localities.
		of the species, well developed large eyes, 11 body segments, and a lobed glabella, broadest in front. It ranges through all countries, and its vertical distribution is very great. Beginning in Lower Bala, it reaches to the upper Devonian—the latest known species being in the Barnstaple or uppermost group.	
Gk	*Phacops* (*Acaste*).	**Phacops, Acaste, Chasmops, Odontochile, Cryphæus,** are subgenera.	
Gk	Pl. 1 G, figs. 17—19, p. 162.	**Phacops** (*Acaste*) **apiculatus,** Salter (Siluria, 4th ed. 1867, p. 69, fig. 14, woodcut, f. 3). A small form, much like the *P. Downingiæ* of the Wenlock rocks.	b. 196, Pwllheli; Bala; Glyn Ceiriog; Conway River; Llanwyddyn; Rhiwargor; Rother Bridge?
Gk	Pl. 1 G, figs. 12—14, p. 159.	**Phacops** (*Acaste*) **alifrons,** Salter (Mon. Brit. Trilob. 1864, pl. 1, figs. 31—34). A peculiar convex form, very like *P. sclerops* of Sweden. The glabella runs out above laterally into the cheeks—hence the name.	b. 198, Capel Garmon; Glyn Diffwys; Bala; Wilfa, Penmachno; Meifod.
Gk	*Odontochile truncato-caudata.* Pl. 1 G, figs. 20, 21, p. 162.	**Phacops** (*Chasmops*) **macroura,** Sjögren (Mon. Brit. Tril. 1864, pl. 4, figs. 18—23). A large trilobite (called Cat's Head trilobite) very common in all the Middle Bala rocks. The peculiar visage of the head is given by the swelling of the upper glabella lobes, and contraction of the lower. [Two or three of these species used to be confounded under the name *P. Odini*, Eichwald].	a. 131, Grug, Llandeilo; Blaen-y-cwm, Nantyr; Coniston; Applethwaite Common.
Gk	*Chasmops Odini?* Pl. 1 G, figs. 22, 23, p. 164.	**Phacops** (*Chasmops*) **conophthalmus,** Bœck.? (Mon. Brit. Tril. 1864, pl. 4, figs. 24, 25). This has a wider glabella and a much shorter tail than the other species.	Alt-yr-Anker, Meifod; a. 128, Llansantfraid, Glyn Ceiriog.
		Calymene, Brongniart. Scarce less widely spread, but not quite so long-lived as *Phacops.* It has a very compact form—but 13 body segments, and a lobed glabella which is smallest in front. The eye was soft, and is seldom preserved. The species are many, and not easy of distinction.	

Case and Column of Drawers.	Reference to McCoy's Synopsis: and Figures of Genera.	Names and References; Observations, &c.	Numbers and Localities.
Gk	*C. brevicapitata*, Pl. 1 F, fig. 6 (not 4, 5), some specimens labelled and catalogued as *C. subdiademata.*	**Calymene Senaria,** Conrad (Salter, Mon. Brit. Trilob. 1865, pl. 9, figs. 5—11, p. 97). One of the very common Bala or Caradoc fossils —appearing to inhabit every kind of sea-bottom. It is often confounded with the *C. Blumenbachii*, var. *brevicapitata,* Portlock.	Bryn Melyn, Bala; Cefn Coedog; Llanwddyn; Alt-yr-Anker, Meifod; Coniston; Horton (very large). a. 135, Pwllheli; High Haume. a. 136, Applethwaite Common.
Gk		**Calymene Blumenbachii,** Brongniart (Sil. Syst. pl. 7, figs. 6, 7. Salter, Mon. Brit. Trilob. p. 93, pl. 8, figs. 7—16). The common Dudley trilobite, in its typical form, with a large broad glabella, is not found in the Bala rocks, but a variety, with narrower glabella, viz.—	
Gk	*C. brevicapitata,* p. 165.	**Calymene,** *var.* **Caractaci,** Salter (Mon. Brit. Tril. p. 96, pl. 9, figs. 3—5. *C. brevicapitata,* Portlock, Geol. Rep. on Tyrone, &c.), is one of the frequent fossils in the Bala beds.	a. 136*, Horderley; above Rother Bridge.
Gk	Pl. 1 G, figs. 24—31, p. 168. 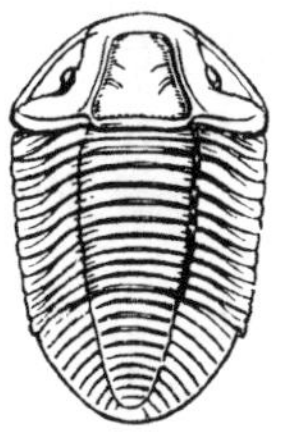	**Homalonotus bisulcatus,** Salter (Mon. Brit. Tril. 1865, pl. 10, figs. 3—10). *Homalonotus,* so named from the uniform scarcely lobed condition of the dorsal surface, is very like Calymene else, and always accompanies it in the old rocks. It however ascends in force into the Devonian, while Calymene does not.	b. 184, Maes Meillion, Bala; Bryn Eithin, Penmachno; Garefawr, Welchpool; a. 111, Pwllheli; Acton Scott, Shropshire (where it is large and abundant).
Gk		**Homalonotus Sedgwickii,** Salter (Mon. Brit. Tril. p. 107, woodcut, fig. 25).	a. 228, Llanwddyn, Montgomeryshire; a. 227, Ravenstonedale.
Gk	Pl. 1 E, fig. 20, p. 168.	**Homalonotus rudis,** Salter (Mon. Brit. Tril. pl. 10, figs. 12—14).	a. 229, Capel Garmon.
Gk	*Isotelus,* p. 169.	**Asaphus Powisii,** Murchison (Salter, Mon. Brit. Tril. pl. 23, figs. 2—5, p. 154). As characteristic of the Middle Bala rocks as the *A. tyrannus* of the Lower Bala, easily distinguished by the faint ribs of the tail.	a. 190, Pen Cerrig Serth, N. of Bala; Bryn Eithin, Penmachno; Bwlch-y-groes; Meifod; Llanwddyn.

Case and Column of Drawers.	Reference to McCoy's Synopsis: and Figures of Genera.	Names and References; Observations, &c.	Numbers and Localities.
Gk	*A. laticostatus,* Pl. 1 E, fig. 18 (not fig. 18 a), which is Lower Bala, p. 170.	**Asaphus radiatus,** Salter (Mon. Brit. Tril. pl. 18, figs. 1—5). Not quite a common fossil in certain parts of the Bala beds—but yet not rare. It seems only to have lived in calcareous mud.	b. 187, Rhiwlas Bala. [The Girvan locality is of May Hill Sandstone.]
Gk	*Dysplanus centrotus?* 173. Pl. 1 E, fig. 19. *Ill. latus,* p. 172. Pl. 1 E, fig. 17.	**Illænus Bowmanni,** Salter (Mon. Brit. Tril. 1867, pl. 28, figs. 6—13). A very common fossil. Unlike most of the *Illæni* it has only 9 body rings like *I. centrotus,* Dalm. It has been often confounded with that spine-headed species.	a. 129, Llanwddyn; Keisley, near Dufton, Westmorland. b. 188, Wrae Quarry, Tweed (*I. latus*).
Gk	Pl. 1 G, fig. 36, p. 171.	**Illænus Davisii,** Salter (Mon. Brit. Tril. p. 194, pl. 29, figs. 10—16). 10 rings to the body, very small eyes.	a. 130, Rhiwlas, Bala; Glyn Diffwys; Llanwddyn.
Gk	Pl. 1 G, figs. 33—35, p. 172.	**Illænus Rosenbergii,** Eichwald (Salter, Mon. Brit. Tril. pl. 29, figs. 2—6, 1867). See discussion in Appendix to Synopsis, p. 4, as *I. Murchisoni.*	a. 141, Coniston; Sunny Brow; Horton in Ribblesdale.
Gk		**Illænus,** small sp. like *I. Barriensis* (probably of the section *Bumastus*), Mon. Brit. Trilob. 1867, p. 215, fig. 57.	b. 189, Mynydd Fron Frys, Chirk.
	Pl. 1 L, fig. 3, p. 336, note.	**Harpes parvulus,** McCoy, Synopsis, p. 336, note.	b. 228, Wrae Quarry, Upper Tweed.
		Mollusca Brachiopoda.	
		The Brachiopods, few in the existing ocean fauna, were of all shells the most prolific in numbers and kinds in the seas of the Cambrian and Silurian æras. Some of the earliest formed genera—*Discina, Lingula* or their allies—have been already quoted from Lower or Middle Cambrian rocks. But the great abundance of *Orthides, Leptænæ,* and such like shelly forms in the Bala rocks give quite a character to the deposit.	

Case and Column of Drawers.	Reference to McCoy's Synopsis: and Figures of Genera.	Names and References; Observations, &c.	Numbers and Localities.
		Lingula, Bruguière. The true *Lingulæ* appear to have commenced existence in the Bala rocks; as the *Lingulellæ* are characteristic of the Middle Cambrian or Lingula Beds. Lingula is the most equivalve, and least complex of all the genera of Brachiopods, the muscles widely separated, the shell is horny and flexible, without calcareous matter or any hinge. The spiral arms are fleshy only. No calcareous spires occur in palæozoic rocks till we reach the Devonian and Carboniferous periods.?	
G 1	*Ling. Davisii,* in part. Pl. 1 L, fig. 6.	**Lingula ovata** *, McCoy (Davidson, Mon. Sil. Brach. p. 38, pl. 2, figs. 19—23). [The specimen figured is a squeezed specimen of *L. Davisii* from the true Lingula flags, p. 39.]	b. 202, Coniston (as *L. Davisii*); Bryn Melyn, Bala, true species; b. 204, Gelli Grin, Bala.
	Pl. 1 L, fig. 8, p. 254.	**Lingula tenuigranulata,** McCoy (Davidson, Sil. Brach. p. 37, pl. 2, figs. 9—14). One of the largest species known in any rock.	a. 151, Alt-yr-Anker, Meifod. (Prof. Sedgwick and Mr Salter).
	p. 253.	**Lingula longissima,** Pander, Beitrage, pl. 3, fig. 21. Dav. Sil. Brach. pl. 3, figs. 28—30.	a. 159, Mynydd Fron Frys, Chirk.
G 1	*Pseudocrania,* p. 187. Pl. 1 H, figs. 1, 2.	**Crania divaricata,** McCoy (Davidson, Mon. Sil. Brach. pl. 8, figs. 7—12). The genus Crania is world-wide from the Bala beds to the present day.	a. 152, Bryn Melyn, Bala; Pont-y-glyn, Diffwys.
G 1	do.	**Trematis** (*Discina*) **corona,** MSS. A giant species. The genus differs from Discina by its marginal foramen for the byssus. *T. punctata* Sow. (Dav. Sil. Brach. pl. 6, fig. 9).	b. 201, Pusgill, Dufton, Westmorland (Prof. Harkness). Horderley.
G 1	*Spirigerina,* p. 197.	**Atrypa marginalis,** Dalm. (Siluria, 2nd ed. pl. 22, fig. 19). A common fossil in beds above the Bala limestone; more rare in N. Wales and Ireland. No hinge line—but calcareous spires placed so that the spire lies flat in the valve—mark this genus.	Pont-y-glyn, Diffwys; Blaen-y-cwm, Nantyr.

* The true *L. ovata,* figured by McCoy from the Bala rocks of S. Ireland, is a common and characteristic fossil of this age.

Case and Column of Drawers.	Reference to McCoy's Synopsis: and Figures of Genera.	Names and References; Observations, &c.	Numbers and Localities.
G 1		**Atrypa** [? **Headi,** Billings. var. **anglica,** Davidson, Sil. Brach. t. 22, figs. 1—8].	Common at Bird's Hill, Llandeilo, &c.
G 1	*Hemithyris depressa,* p. 201.	**Triplesia** ? **Maccoyana,** Davidson, Sil. Brach. p. 199, pl. 24, f. 29 (*R. depressa,* McCoy, not of Sowerby in Silurian System). In Davidson's Monog. p. 123, *R. depressa,* Sow. is an Athyris.	a. 157, Brynbedwog, near Bala.
G 1		**Rhynchonella** (or *Atrypa*) sp.	b. 206*, Ravenstonedale.
G 1	*Hemithyris,* p. 199, Pl. 1 H, figs. 6—8.	**Meristella angustifrons,** McCoy (Siluria, 3rd ed. woodcut 49, fig. 2). A common fossil in Llandovery rocks, and very rare in Middle Bala.	b. 206, Craig Head, Girvan.
G 1		**Rhynchonella,** sp. (*Headi,* Billings?)	Coniston (same at Grug, Llandeilo).
G 1	*Meristella. Hemithyris.* Pl. 1 L, fig. 5, p. 203.	**Rhynchonella nasuta,** McCoy sp. (Davidson, Mon. Sil. Brach. p. 173, pl. 23, f. 19).	a. 156, Craig Head, Girvan.
G 1		**Rhynchonella,** sp.	Yspytty Evan.
		Several other specimens undescribed are common in Wales, Ireland, Scotland.	
G 1	*Pentamerus lens,* p. 209.	**Pentamerus lens,** Sow. sp. Sil. Syst. 21, f. 3.	b. 207, Coniston.
		New genus of **Rhynchonellidæ,** long mesial septum in the dorsal valve (looks very like a *Pentamerus* at first sight).	
	p. 208.	[*P. globosus,* supposed to be found at Beaver's Grove, near Conway, is really a Bellerophon, and must not figure here. No species of *Pentamerus* has ever occurred in Lower or Middle Bala !]	
	p. 212.		
G 1		**Porambonites intercedens,** Pander (*Spirifer porambonites,* Geol. Russ. Vol. II. p. 131, pl. 2, fig. 3). A very peculiar genus, allied to Orthis; but with a curiously punctate surface.	a. 205, Wrae Quarry, Upper Tweed.

Case and Column of Drawers.	Reference to McCoy's Synopsis: and Figures of Genera.	Names and References; Observations, &c.	Numbers and Localities.
		Orthis, Dalman. A genus of all others characteristic of the great Bala and Silurian groups. Less abundant in Devonian, and not quite extinct in Carboniferous times. It is easily distinguished by its convex form, and single cardinal process, from *Leptæna* or *Strophomena*, which genera occur with it, and of large size.	
G 1	p. 213.	**Orthis Actoniæ,** Sowerby (Siluria, 2nd ed. pl. 5, fig. 11), and Salter, Mem. Geol. Surv. III. p. 339, pl. 21, figs. 1—8. A shell easily distinguished from its associate *O. flabellulum*, by having the opposite valve convex; the concave one being the dorsal one.	Bryn Eithin, Penmachno. Gelligrin; Bryn Melyn, &c., near Bala; Cader Dinmael; Pont-y-glyn, Diffwys; Blaen-y-cwm, Nantyr; Llansantfraid, Glyn Ceiriog; Alt-yr-Anker, Meifod; Tyn-y-Cabled, Llanfyllin. a. 149, Acton Scott, Shropshire; Ingleton; Thornton; Horton; High Haume.
G 1	*O. callactis*, p. 214.	**Orthis,** var. (as *O. callactis*, but not of Dalman). It is only a large variety of *O. Actoniæ*.	Gaerfawr, Welchpool; Bala. One very large variety at Wrae Quarry, Up. Tweed.
G 1	p. 213.	**Orthis biloba,** Linn. (Siluria, 2nd ed. pl. 20, fig. 14). Not uncommon in Bala rocks. Swarms in Wenlock.	Cefn Goch, Glyn Ceiriog.
G 1	p. 214.	**Orthis calligramma,** Dalman (Salter, in Mem. Geol. Surv. III. p. 335, pl. 22, fig. 1). Both valves are equally convex in all the varieties.	Llansantfraid, Glyn Ceiriog. [The typical form of the species is of Lower Bala age in Britain.]
G 1	*O. plicata*, p. 214.	**Orthis,** id. var. **plicata,** Salter (*O. plicata*, Silurian System, pl. 21, fig. 6). *O. calligramma*, var. *plicata*, Salter, Mem. Geol. Surv. III. p. 336, pl. 22, fig. 5).	Gelligrin; b. 210, Bryn Melyn, Bala; Cwm of the Cymmerig, Bala; Llansantfraid, Glyn Ceiriog; Bryn Evan, Penmachno; Pwllheli; Cader Dinmael, near Meifod, abundant. Gaerfawr, Welchpool; Das Eithin; Horton in Ribblesdale.

Case and Column of Drawers.	Reference to McCoy's Synopsis: and Figures of Genera.	Names and References; Observations, &c.	Numbers and Localities.
G 1	*O. flabellulum*, p. 219.	**Orthis**, id. var. **virgata**, Salter, Mem. Geol. Surv. III. p. 336, pl. 22, fig. 3 (*O. virgata*, Sil. Syst. pl. 20, fig. 15), Siluria, 2nd ed. pl. 5, fig. 9.	b. 215, Coniston; b. 216, Applethwaite; Gelligrin, Bala; Alt-yr-Anker, Meifod, Montgomeryshire (as *O. rigida*).
G 1	*O.* var. *a. calliptycha*, p. 215.	**Orthis**, id. var. **calliptycha**, McCoy. A pretty variety with coarse ribs; the interspaces strongly ribbed with wire-like striæ.	b. 214, Llansantfraid, Glyn Ceiriog.
G 1	p. 214. *O. rigida*, p. 226.	**Orthis**, id. var. **Wallsalliensis**, Salter (*O. Wallsalliensis*, Davidson. See Mem. Geol. Surv. Vol. III. p. 337, pl. 22, figs. 6, 7.	b. 212, Rhiwargor (more common in Llandovery rocks). Gelligrin; Trawscoed; b. 211, Gaerfawr (as *O. confinis*). b. 213, Gaerfawr, Welchpool.
G 1	p. 216.	**Orthis**, id. var. **crispa**, McCoy (Sil. Foss. Irel. p. 29, t. 3, fig. 10. A remarkable small species, with strong waves and ridges of growth decussating the ribs.	a. 158*, Bala; Blaen-y-cwm, Nantyr; a. 158, Helmsgill, Dent.
G 1	p. 219.	**Orthis flabellulum**, Sow. (Salter, Mem. Geol. Surv. III. p. 338, pl. 21, figs. 9—16). A handsome Bala shell, with fan-like ribs 20 to 30.	a. 150, Bodean, Pwllheli, Bettws; Moel-y-garnedd, W. of Bala; Gelligrin; Cader Dinmael; Blaen-y-cwm, Nantyr; Snowdon top; Llangedwyn; Llanwddyn; Meifod; Moel Uchlas.
G 1	*O. sarmentosa.* Pl. 1 H, figs. 25—28, p. 227.	**Orthis flabellulum**, var. finest ribs (Salter, Mem. Geol. Surv. III. pl. 21, fig. 17), looks like a distinct species, but is easily traceable into *O. flabellulum.* N.B. *O. sarmentosa*, McCoy, Sil. Foss. Ireland, is the *O. testudinaria*, distorted by cleavage.	b. 219, Llyn Ogwen, abundant.
G 1	Pl. 1 H, figs. 41, 42, p. 223.	**Orthis porcata**, McCoy (*O. occidentalis*, *O. sinuata*, &c. of Hall. Salter, Mem. Geol. Surv. Vol. III. p. 338, pl. 19, fig. 4). Very like extreme varieties of *O. flabellulum*, from which, when fine-ribbed varieties of both are compared, it is extremely difficult to separate it. The interior characters are alike in both.	b. 221, Coniston; High Haume; Dalton; a. 155, Horderley; Corwen; Bala; Llansantfraid, Glyn Ceiriog; Meifod; a. 154, Alt-y-gader; Alt-yr-Anker; Llanfyllin (Pen-y-Park), Corwen. Common in all varieties, N. Ireland.

Case and Column of Drawers.	Reference to McCoy's Synopsis: and Figures of Genera.	Names and References; Observations, &c.	Numbers and Localities.
G1	Pl. 1 H, figs. 41, 42, p. 223.	**Orthis,** sp. near to *O. porcata.*	Mynydd-y-gaer, Llanefydd.
G1	p. 216.	**Orthis elegantula,** Dalm. (Siluria, 2nd ed. pl. 5, fig. 5). The weed of the Silurian and Upper Cambrian rocks; appearing under many varieties in Britain and Sweden; apparently unknown in S. Europe.	Everywhere in the Bala rocks; in Caernarvonshire at Pwllheli; Snowdon; Moel Hebog, &c. All round Bala Lake in Merionethshire. The Berwyns, east side (Milltir Cerrig), Welchpool; Meifod; Moel Uchlas near Llanfyllin; Horderley; Acton Scott; and all the Caradoc district.
G1	p. 221.	**Orthis parva,** Pander. Considered a distinct species by De Verneuil and McCoy, and others. I think it only a sharply ribbed variety of *O. elegantula,* with which it is generally intermixed.	Dinas Mowddwy; Llansantfraid, Glyn Ceiriog; Bryn Melyn, Bala; Yspytty Evan. Moel Uchlas, &c. Ardwell, Girvan, Ayrshire.
G1	p. 221.	**Orthis parva,** var. **avellana,** De Verneuil. Geol. Russ. t. 13, figs. 3, 4.	b. 217, N. of Tremadoc; Tan-y-Bwlch-y-groes.
G	*Spirifera,* p. 192.	**Orthis biforata,** Schloth. (Davidson, Mon. Brach. Pal. Soc. Introd. pl. 8, fig. 146). One of the commonest of the Bala types, ranging from Lower Bala to Wenlock rocks. It looks like a Spirifer, and was so named by Prof. McCoy, as it had been by other authors. There are no internal shelly spires. It is easily divisible into —	
G	p. 192.	**Orthis biforata,** proper, with 6 to 9 ribs on the raised front.	b. 239*, Troutbeck; b. 239, Bryn Melyn, Bala.
G	*Spirifera,* p. 192.	**Orthis biforata,** var. α *Lynx* (*Spirifer,* Eichwald &c.), 4 ribs on the fold.	a. 145, Meifod; Llanfyllin; Bala; Glyn Diffwys, Corwen; High Haume, Dalton; &c.
G	p. 192.	**Orthis biforata,** var. β *dentatus.*	Coniston; Rhiwargor; Llanwddyn; Cyrn-y-Brain, Wrexham; Coniston; Ravenstonedale.
G1	*Spirifera,* p. 193.	**Orthis fissicostata,** seems quite distinct. **Orthis** (*O. biforata,* var. *fissicostatus*), McCoy.	b. 235, Tyn-y-Cabled; b. 236, Alt-yr-Anker; b. 237, Penarth, Meifod; Ravenstonedale.

Case and Column of Drawers.	Reference to McCoy's Synopsis: and Figures of Genera.	Names and References; Observations, &c.	Numbers and Localities.
G	p. 193.	**Orthis,** new sp. closely allied to *O. biforata.*	b. 233, Bala; b. 234, Nant-yr; a. 241, Tyn-y-Cabled.
G l	*O. retrorsistria.* Pl. 1 H, fig. 12, p. 224, and note, p. 217.	**Orthis alternata,** Sow. (Siluria, 2nd ed. pl. 6, fig. 5). There can be no doubt the Welch fossil is the dwarf form of this common Horderley species. It is so easily distinguished by the interior cast, and so hardly distinguishable by the exterior, that it has been confounded constantly with *Strophomena alternata* in British Works (the Synopsis included).	a. 144, Cerrig-y-Druidion, in millions. Penmachno; Cernioge; Corwen; Bwlch Llandrillo; Cefn-y-Coedog; Llangedwyn; Das Eithin, Hirnant, and Llanwddyn, Montgomeryshire; Pen Cerrig Serth, N. of Bala; Glyn Diffwys; Meifod; Llangfyllin (the Carn Goran species is distinct and a sp. of *Strophomena;* see Mem. Geol. Surv. III. p. 340).
G	*Spirifera,* p. 194.	**Orthis insularis,** Eichwald (De Vern. Geol. Russia, t. 8, fig. 7). One of the rare group of smooth Orthides which occur in Bala or Caradoc rocks, over the N. Hemisphere at least. A very similar species is found in India. The dorsal valve is highly convex—the ventral, concave.	b. 241*, Gelligrin, Bala; b. 241, Coniston.
G m	*Leptæna spiriferoides,* p. 246.	**Orthis Spiriferoides,** McCoy (Siluria, 3rd ed. woodcut 37, p. 194). Of the same group with *O. insularis,* from which its strong striæ or ribs easily distinguish it.	a. 146, Alt-yr-Anker; b. 251, Gaerfawr, Welchpool; Llanfyllin; Bala; Horderley (and a hundred other Caradoc localities).
G l		**Orthis simplex,** McCoy (Sil. Foss. Ireland, pl. 3, fig. 18). Very like *O. calligramma,* and possibly a variety.	Gelligrin; Llansantfraid.
G l	p. 224.	**Orthis protensa,** Sow. (Siluria, 2nd ed. pl. 9, fig. 22) includes *O. lata.* Sow. Sil. System, t. 22, figs. 8, 9. This common Llandovery species is very doubtfully recognized in the above locality.	Cader Dinmael, Holyhead road, Denbyshire.
G l	Pl. 1 H, figs. 25—28, p. 227.	**Orthis sarmentosa,** McCoy, which is *Orthis testudinaria* in a crushed state.	

Case and Column of Drawers.	Reference to McCoy's Synopsis: and Figures of Genera.	Names and References; Observations, &c.	Numbers and Localities.
G l	p. 228.	**Orthis testudinaria,** Dalman (Siluria, 2nd ed. pl. 5, figs. 1—2). A much coarser striated shell than the *O. elegantula,* and with different hinge-characters, else not unlike that species. *O. testudinaria* is however wholly a Bala rock fossil.	All through the sandy part of the Bala series abundant; Alt-yr-Anker, Meifod; Pwllheli; the Hollies; b. 225, Blaen-y-cwm; Llansantfraid; Glyn Ceiriog; a. 148, Horderley.
G l	Pl. 1 H, figs. 20—24, p. 229.	**Orthis turgida,** McCoy. One of the most convex of species. It is a very rare one. The space between the great muscles of the dorsal valve usually occupied by a low ridge is here a sharp one.	a. 181, N. of Conway.
G l	p. 230.	**Orthis Vespertilio,** Sow. (Siluria, 2nd ed. pl. 6, figs. 1—3). A local British species—so strongly divided into two convex lobes, that it was originally called *bilobata* by Sowerby. The dorsal valve has a sharp keel all down it.	b. 252, Blaen-y-cwm, Nantyr; Bala; Llangollen; Meifod; Welchpool; Bwlch-y-ciban, Montgomeryshire; Horderley; Coniston; Rother Bridge.
G m	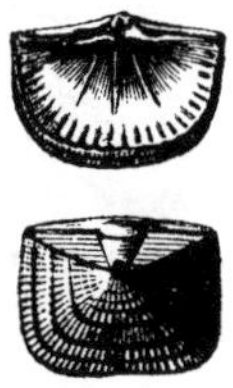	**Orthisina.** Distinguished from *Orthis* no less by the very large area and covered deltidium, than by the peculiar shape of the teeth in the smaller valve.	
G m	p. 231.	**Orthisina adscendens,** Pander (Geol. Russ. Vol. II. p. 203, pl. 12, fig. 3). Striated roughly and with decussating striæ of growth.	b. 258, Cyrn-y-Brain, Wrexham. Llansantfraid, Glyn Ceiriog; a. 142, Cefn Coedog.
G m	Pl. 1 H, fig. 29, p. 232.	**Orthisina scotica,** McCoy. Ribbed strongly in a radiate fashion.	a. 143, Craig Head, Girvan; one specimen, b. 260, said to be from Colmonel, i.e. Lower Bala.
G m	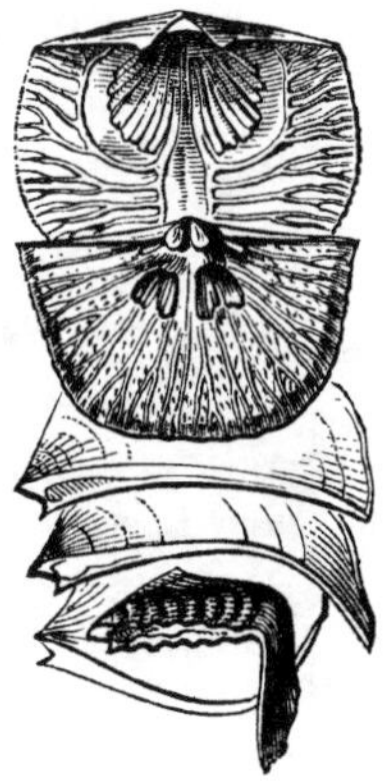 *Leptæna* (McCoy does not admit *Strophomena*).	**Strophomena,** Rafinesque. Flattened shells expanded, not convex as Orthis; one valve frequently concave—the other convex. And the large central tooth or cardinal process double.	Range, Upper Cambrian, i.e. Lower Bala, and Arenig rocks. Devonian.
G m	*Leptæna,* p. 241.	**Strophomena antiquata,** Sow. (Siluria, 2nd ed. pl. 20, fig. 18). A very unusual form of this genus, for it is covered with thick blunt ribs and ridges. The species ranges up into the Ludlow rocks.	b. 254, Coniston, Bala; Llangollen; b. 253, Blaen-y-cwm Nantyr; Llansantfraid.

Case and Column of Drawers.	Reference to McCoy's Synopsis: and Figures of Genera.	Names and References; Observations, &c.	Numbers and Localities.
G m	*Leptæna*, p. 233.	**Strophomena alternata,** Conrad (Hall, Pal. N. York, I. pl. 31 a, fig. 1). Common. A difficult species to recognize as distinct from flatter varieties of *L. deltoidea.*	Girvan, Ayrshire.
G m	do.	**Strophomena.** Like *S. alternata.*	Coniston limestone (limestone nodules in Old Red), Holbeck Gill.
G m	*Leptæna compressa*, p. 242.	**Strophomena concentrica,** *Orthis expansa*, Portl. (Geol. Rep. pl. 37, fig. 1). Distinguished from *S. expansa* by the strong interior concentric ridges, and the teeth also diverge differently.	b. 254*, Tynant, Horderley.
G m	*Leptæna*, p. 233.	**Strophomena corrugata,** Portl. (Geol. Rep. pl. 32, fig. 17). This pretty group of sharply waved and striated species is European, not American. It probably contains many similar species.	Keisley, Dufton (in green slates).
G m	*Leptagonia*, McCoy, p. 248.	**Strophomena depressa,** Dalman (*L. rhomboidalis*, Wahl.), Min. Conch. t. 459. The commonest Brachiopod next to *Orthis elegantula*, in all the slate rocks—ranging from Lower Bala to Lower Carboniferous!	Selattyn Road, Oswestry; Corwen; Cefn Coedog; Llansantfraid, Glyn Ceiriog; and all through N. and S. Wales. Coniston limestone, everywhere.
G m	p. 249.	**Strophomena,** var. γ *ptychotis*, McCoy. Sil. Syst. Ireland.	b. 242, Wilfa Penmachno; Cefn Coedog, Corwen; Llanbedrog; Pwllheli; Cader Dinmael, Corwen.
G m	*Lept.* p. 234	**Strophomena deltoidea,** Conrad. Hall, Pal. N. York, t. 13 a, fig. 3. Under this name Prof. McCoy seems to include some varieties of the *L. tenuistriata*, Sow. and also other species. *L. deltoidea* is well figured by Hall.	Coniston (two species here); Alt-yr-Anker, Meifod.
G m	Pl. 1 H, figs. 38, 39, p. 234. As a var. of *deltoïdea.*	**Strophomena,** var. β *undata* (see *deltoidea*). A marked form, resembling *S. depressa*, but with no recurved portion.	Grug; b. 248, Llandeilo; a. 153, Bala; a. 153*, Cyrn-y-Brain, Wrexham; b. 247, Bryn Melyn, Bala; Glyn Diffwys; Alt-yr-Anker, Meifod.
G m	Pl. 1 H, figs. 36, 37. *Leptæna*, p. 249.	**Strophomena ungula,** McCoy.	Llansantfraid, Glyn Ceiriog; Selattyn Road.

Case and Column of Drawers.	Reference to McCoy's Synopsis: and Figures of Genera.	Names and References; Observations, &c.	Numbers and Localities.
G m	*Stroph. compressa,* p. 242.	**Strophomena expansa,** Sow. (Siluria, 2nd ed. pl. 6, fig. 4). The remarkable striated fan-shaped muscle scars easily distinguish this finely striated species.	Llanfyllin; Meifod; Welchpool; Bala; Penmachno; Pwllheli, and all through the Snowdon district and Shropshire.
		Strophomena tenuistriata, Sow. (Siluria, 2nd ed. pl. 5, fig. 15).	Alt-yr-Anker, Meifod.
G m	*Lept. deltoidea,* McCoy.	**Strophomena bipartita,** Salter (Quart. Journ. Geol. Soc. Vol. 10, p. 74). Must be distinguished by its strong central rib and flatter form from *S. deltoidea.*	b. 246, Acton Scott.
	Pl. 1 H. figs. 33—35, p. 246.	**Strophomena simulans,** McCoy. Rather a doubtful fossil.	b. 250, Blaen - y - cwm, Nantyr; b. 249, Cefn Goch, Glyn Ceiriog. b. 251, Golden Grove? wholly doubtful, both species and locality.
	p. 244.	**Strophomena grandis,** Sow. (Siluria, 2nd ed. pl. 6, figs. 6, 7). Our largest Cambrian *Strophomena,* with minute teeth, a very thin striated shell.	b. 256, Horderley; Bodean, Pwllheli.
	S. grandis, p. 245.	**Strophomena,** sp. (coarser striæ than *S. grandis*).	b. 255*, Cefn Goch, Glyn Ceiriog; b. 256*, Ravenstonedale; Cefn Coedog, near Corwen.
	p. 240.	**Strophomena pecten,** Linn. (Davidson, Introd. Brach. Pal. Mon. pl. 8, figs. 163, 164).	Coniston; Applethwaite Common; Ravenstonedale; Horton; Blaen - y - Cwm, Nantyr, Glyn Ceiriog.
	Leptœna, p. 244.	**Strophomena funiculata,** *Orthis,* McCoy (Sil. Foss. Ireland, p. 30, t. 3, fig. 11).	Cyrn-y-Brain? Llandovery rock.
		Leptæna, Dalman. Composed of those thin (usually transverse) Brachiopods which have the one valve involute on the other.	Range— Upper Cambrian —Silurian.
	p. 240.	**Leptæna transversalis,** Dalm. (Siluria, 2nd ed. pl. 9, fig. 17). Differs from *L. sericea* in the strong internal muscles which are longer, and also in the very spinose inside, the spines leaving deep pits on the cast.	b. 255, Coniston; above Rother Bridge; Llansantfraid, Glyn Ceiriog.

Case and Column of Drawers.	Reference to McCoy's Synopsis: and Figures of Genera.	Names and References; Observations, &c.	Numbers and Localities.
	L. quinquecostata. Pl. 1 H, fig. 31, 32.	**Leptæna transversalis,** var. *undulata,* Salter, Mem. Geol. Surv. III. p. 276. Catal. Mus. Pract. Geol. p. 9, often mistaken for *L. quinquecostata,* but differs in its undulated surface, and many, instead of 3—5 strong ribs.	
	p. 237.	**Leptæna sericea,** Sow. (Siluria, 2nd ed. pl. 5, fig. 14). Extremely common. It has a fine even surface closely striated. The muscular scars are short for the genus.	Every possible locality in Bala rocks both Middle and Upper, N. and S. Wales. Pont-y-Glyn, Diffwys, Corwen; Glyn Ceiriog; Penmachno; Capel Currig; Bettws; Pwllheli; Llanfyllin; Llanfechan; Gaerfawr, Guilsfield, &c., Horderley.
	p. 239.	**Leptæna sericea,** var. *α rhombica,* McCoy, probably not of Phillips.	b. 256, Gelligrin; Bala; a. 242, Alt-yr-Anker, Meifod.
	do.	**Leptæna,** do. (internal cast, quoted in p. 237).	b. 254, Llandrillo.
	Pl. 1 H, fig. 44, p. 239.	**Leptæna tenuissimestriata** is a bad species, made up of *L. sericea* and *L. transversalis.* See Lower Bala group.	
	do.	**Leptæna,** new sp. fine striæ, prominent beak.	Bower Bank, Dent.
	Pl. 1 H, fig. 30, p. 236.	**Leptæna quinquecostata,** *Orthis,* McCoy (Sil. Foss. Ireland, pl. 3, fig. 8).	b. 253, Glenquhaple, 3½ miles S. of Girvan.
	Pl. 1 H, fig. 40, p. 239. 	**Leptæna tenuicincta,** McCoy (An. Nat. Hist. 2 ser. 8, p. 401). A species much longer than wide. Very common in S. Wales (Bala limestone) with large long muscles.	b. 252, Cefn Grugos, Llanfyllin; Keisley, Dufton (Prof. Harkness).
		Lamellibranchiata or Conchifera.	
	Ordinary Bivalve shells.	(*Pleuroconcha—Acephala,* of various authors).	
		Prof. Phillips has suggested (Mem. Geol. Surv. Vol. II. Pt. 1, p. 263—275), that the earliest *Lamellibranchiata* (not Brachiopods) known, and these do not descend further than the Arenig or Skiddaw group, are everywhere referable to the *Modiola* and *Arca* groups, notably to the latter. These groups stand at the junction, so to speak, of the *Dimyaria* and *Monomyaria* groups. The fact is worthy of notice, in reference to any theory of development of organic forms.	

Case and Column of Drawers.	Reference to McCoy's Synopsis: and Figures of Genera.	Names and References: Observations, &c.	Numbers and Localities.
G m	Pl. 1 I, figs. 1, 2, p. 261.	**Pterinea pleuroptera,** Conrad?	a. 166*, Cyrn-y-Brain, Wrexham.
	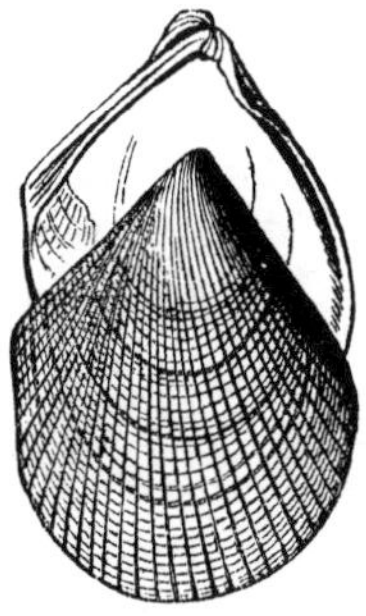	**Ambonychia costata,** Conrad.	Not in the collection.
G m	*Ambonychia.*	**Modiolopsis,** McCoy. Confessedly a genus of convenience, it being extremely unlikely that *Modiola*-like shells, so different in external characters from the living forms, should have possessed no internal characters whereby to separate them. (The front muscular scars are deep and strongly defined, however.) Common in Bala rocks.	a. 160, Cader Dinmael.
G m	Pl. 1 I, figs. 17, 18, p. 267.	**Modiolopsis modiolaris,** Conrad sp.	a. 160, Cader Dinmael. a. 160*, Horderley.
G m	*Avicula?* p. 258.	**Modiolopsis orbicularis,** Sow. (*Avicula,* Sil. System, t. 20, fig. 3, Siluria, 2nd ed. pl. 7, fig. 1). A very round and rather long than wide form.	a. 166, Acton Scott, Shropshire; and abundant in all sandy Bala rocks (rises into May Hill sandstone). J. W. S.
G m	Pl. 1 I, fig. 16, p. 266.	**Modiolopsis inflata,** McCoy. (*M. antiquata* of McCoy is a mistake. The species is Wenlock, and does not occur in Bala rocks.)	a. 208, Pen-Cerrig, Serth. N. of Bala.
G m	Pl. 1 K, fig. 17, p. 272.	**Lyrodesma plana,** McCoy. A genus with the characteristic numerous teeth of the *Arcacidæ,* but disposed fan fashion under the beak.	a. 209, Yspytty Evan.
G m	Pl. 1 K, figs. 7, 8, p. 273.	**Cleidophorus? ovalis,** McCoy. Doubtfully referable to this genus, which was proposed by Hall for toothless shells like flattened *Cucullellæ.*	a. 210, Dolydd Ceiriog waterfall, in the Berwyn Mountains.
	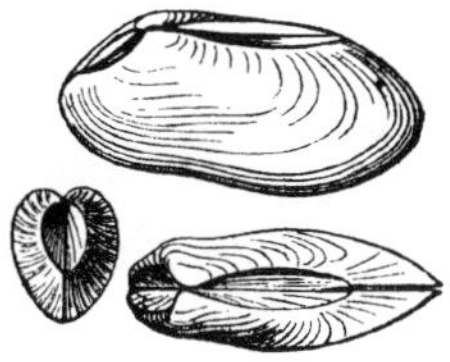	**Orthonotus,** Conrad. A convenient group, intended to enclose a number of Mytiloid shells, thin, without teeth; with a form like *Myacites* (but certainly with no relation to it), no pallial sinus, no wrinkled or granulated epidermis; and with dorsal and anterior lunettes.	
G m	Pl. 1 I, fig. 23, p. 275.	**Orthonotus nasutus,** Conrad sp.	a. 211, Horderley.

Case and Column of Drawers.	Reference to McCoy's Synopsis: and Figures of Genera.	Names and References: Observations, &c.	Numbers and Localities.
G m	*Arca*, Pl. 1 K, figs. 2, 3, p. 283.	**Ctenodonta Edmondiiformis,** McCoy.	a. 212, Alt-y-Gader.
G m	*Nucula levata*, Pl. 1 K, figs. 4, 5, p. 285.	**Ctenodonta varicosa,** Salter, Siluria, 3rd ed. woodcut 39, fig. 4. Mem. Geol. Surv. III. p. 345, woodcut 13, fig. 1, p. 343.	a. 164, Milltir Cerrig (milestone), on the Llanwddyn road. The species is a common one at Conway falls.
G m	p. 284.	**Cucullella antiqua?** Sow. (Siluria, 2nd ed. pl. 34, fig. 16).	a. 213, Conway falls.
		Palæarca, Hall (*Cyrtodonta*, Billings. *Cypricardites*, Hall). With the general structure of Cucullæa these ancient bivalves want the posterior internal plate, and are closely related to Ctenodonta, and have affinity with *Pterinea*. Several species are known at Bala, and should be added to the collection.	
	Sea Butterflies generally of large size.	**PTEROPODA AND HETEROPODA (or Nucleobranchiata).**	
		The Pteropods are among the Mollusca the most simple in structure of the cephalate orders. They are also the most ancient. The *Heteropoda* or Nucleobranchs are low down in the scale, being nearest like the embryos of Gasteropods. Both Pteropoda and Heteropoda grew to an enormous size in the older Palæozoic times.	
		Theca, Morris, 1844. A triangular *Pteropod* shell, with an operculum (Barrande). The genus is related to the much smaller living genus *Creseis*, Rang. These were floaters, and as such the genus had a very wide range in olden time. Such fossils too, not being subjected to the full force of local elevation, have been usually long lived. *Theca* is found all through the Lower and Upper Cambrian to the Devonian epoch.	
G m		**Theca reversa,** Salter (*T. Forbesii*, Hall, not of Sharpe, figured in Siluria, 3rd ed. p. 199, Foss. 41, fig. 1). Mem. Geol. Surv. Vol. III. p. 353 (woodcut 14, fig. 6, p. 347).	Common in Wales and Shropshire.

Case and Column of Drawers.	Reference to McCoy's Synopsis: and Figures of Genera.	Names and References; Observations, &c.	Numbers and Localities.
		Conularia, Miller, 1818. A large horny, probably transparent shell, highly ornamented, like certain species of *Cleodora*. The genus has a long range from the Tremadoc rocks to the Coal measures! Some of the species had long lobes between the four angles converging in age so as to close in the aperture.	
G m	*C. cancellata*, p. 287.	**Conularia Sowerbyi,** Defr. (Siluria, 2nd ed. pl. 25, fig. 10). Ranges from Bala rocks to Ludlow,—compressed.	a. 214, Bryn Melyn, Bala. The species is easily distinguished by its compressed form, and not square section, from the coal measure fossil, *Con. quadrisulcata*.
	Heteropods—such as *Atlanta*.	**Bellerophon,** Montfort, 1808 (*Bucania*, Hall, and *Euphemus*, McCoy).	
		Most authors have agreed to regard Bellerophon as a Heteropod or Nucleobranch; of larger size and stronger shell than the living forms. In support of this view we have the facts, that many species have extremely thin and ornamented forms of shells; that they have a deep notch in the mouth like the living Atlanta; and globose forms in others, like the majority of the small living (Heteropods) Nucleobranchs. Against the idea that they could be single chambered Cephalopods, we have 1st, their thin and small shells (few attain the size of the smallest shelled Cephalopods); 2nd, if the analogy be drawn with *Argonauta* (instead of *Carinaria* as usual), it will be necessary to suppose that Decapod cuttlefishes abounded in Cambrian times, an idea wholly at variance with the observed facts, which limit Decapods to the secondary and tertiary epochs. I cannot follow Prof. McCoy here.	
G m	p. 308.	**Bellerophon bilobatus,** Sow. (Siluria, 2nd ed. pl. 7, fig. 9). The surface is covered with an extremely fine reticulation, not yet figured in any work. This kind of ornament is conspicuous and common in the genus.	a. 161*, Dolydd Ceiriog Waterfall, in the Berwyns. a. 161, Teirw River, N. Wales, S. of Llangollen; Horderley and Cheney Longville (in Shropshire), most abundant; Dinas Mowddwy; Llanwddyn.

Case and Column of Drawers.	Reference to McCoy's Synopsis: and Figures of Genera.	Names and References; Observations, &c.	Numbers and Localities.
Gm	Pl. 1 L, fig. 25, p. 311.	**Bellerophon subdecussatus,** McCoy. A form belonging to the subgenus *Bucania,* which has the whorls exposed in the umbilicus.	a. 163*, Alt-yr-Anker, Meifod.
G	As *B. ornatus,* p. 310.	**Bellerophon nodosus,** Salter (*B. ornatus,* McCoy, not of Conrad. Mem. Geol. Surv. III. p. 349, woodcut 15).	a. 163, Teirw River, S. of Llangollen.
	B. expansa, p. 309.	(**Bellerophon globatus**? McCoy is not worth inserting. The original species was named from an obscure broken *B. expansus,* Sow.) N.B. *Maclurea macromphala,* McCoy, should be placed here, for *Maclurea* is a Heteropod. It is probable however that this shell is not a Maclurea.	A Ludlow species.
		Gasteropoda.	
case G	*Univalve shells, spiral shells.*	Only the herbivorous genera, or such as by their round mouths appear to have been such, are found in the old slate rocks, *Pleurotomaria, Murchisonia,* and *Holopea,* &c. are probably all related to the *Ianthinidæ* and *Litorinidæ,* vegetable feeders.	
G m	*Pleurotomaria,* p. 291.	**Murchisonia turrita,** Portl. sp. (Geol. Report, p. 413, pl. 30, fig. 7?). Rather doubtfully identified. N.B. *Pleurotomaria, Murchisonia,* and *Hormotoma,* &c. include a host of species very like one another in habit, and living in crowds together, probably much in the way of the violet snail (*Ianthina*). It is difficult to distinguish the species, unless close attention be paid to bands, ridges, &c. &c. The beaded-whorled species are distinguished as *Hormotoma* (see fig.), and these, which are just as common as the others, form perhaps a more natural group than either *Murchisonia* proper or *Pleurotomaria.* Of the latter genera, only the shorter forms are called *Pleurotomaria,* but these are not so *Trochus-shaped* as the Oolitic species.	a. 167*, Cyrn-y-Brain, Wrexham; Llyn Ogwen. Bala.

Case and Column of Drawers.	Reference to McCoy's Synopsis: and Figures of Genera.	Names and References; Observations, &c.	Numbers and Localities.
G m	Pl. 1 L, fig. 20, p. 292.	**Murchisonia cancellatula,** McCoy.	a. 174*, Alt-yr-Anker, Meifod.
	Pl. 1 K, fig. 4, p. 294. (not 1 L, fig. 21).	**Murchisonia simplex,** McCoy. Distinguished from the next by a blunt median keel, and one above and below it.	Glenquhaple; a. 167, Alt-yr-Anker.
G m	Pl. 1 K, fig. 43, p. 293.	**Murchisonia gyrogonia,** McCoy (*M. scalaris*, Salter, but this was named from very imperfect specimens).	a. 174, Yspytty Evan. Llanfechan.
G m	Pl. 1 K, fig. 42, p. 294.	**Murchisonia pulchra,** McCoy (Sil. Foss. Irel. pl. 1, fig. 19). Perhaps a distinct species from the Irish one, which is from Llandovery rocks.	Alt-yr-Anker; a. 215*, N. of Tremadoc.
G m	Pl. 1 L, fig. 23, p. 298. *Euomphalus.* 	**Cyclonema lyrata,** McCoy. *Cyclonema* is a genus of Prof. Hall's, intended for thin shells of the *Litorina* group. These and all such shells were formerly referred to *Turbo.*	a. 179*, Llansantfraid, Glyn Ceiriog.
G m	Pl. 1 K, fig. 36, p. 296. *Turbo crebristria.*	**Cyclonema crebristria,** McCoy. (*Turbo* and *Holopea*, McCoy). Mem. Geol. Surv. III. p. 347, woodcut 14, fig. 5). A thin shell, spirally striate, ranges to May Hill Sandstone.	a. 173, Bala (Gelli Grin); Alt-yr-Anker, Meifod; Mynydd Fron Frys, Chirk.
G m	Pl. 1 K, figs. 37, 38, p. 299. *Euomphalus.*	**Cyclonema triporcata,** McCoy (possibly referable to the genus *Trochonema*).	a. 173*, Cyrn-y-Brain, Wrexham.
G m		**Holopea,** Hall. A genus of smooth or nearly smooth shells, like *Litorina:* but evidently not operculate as Turbo. It includes a great many of the ordinary univalves in Cambrian and Silurian rocks.	
G m	Pl. 1 K, fig. 41, p. 296. *Trochus constrictus.*	**Holopea Striatella,** Sow. (Siluria, 2nd ed. pl. 7, fig. 4).	a. 169*, Bryn Melyn, Bala; a. 169, Cymmerig Brook, do.
G m		**Holopea conica, exserta, lymnæoides,** &c. occur at Rhiwlas near Bala. (Mem. Geol. Surv. III. woodcut 14, p. 347.) These shells were referred by Forbes to such living types as the floating *Litiopa.* (See Mem. Geol. Surv. Vol. III. p. 346.)	
G m		**Raphistoma,** Hall.	

Case and Column of Drawers.	Reference to McCoy's Synopsis: and Figures of Genera.	Names and References; Observations, &c.	Numbers and Localities.
G m	p. 291. (*Pleurotomaria lenticularis.*)	**Raphistoma æqualis,** Salter (Siluria, 3rd ed. Foss. 40, fig. 2). It is found that the proportions of the whorls in this species differ from those of the May Hill Sandstone form.	Alt-yr-Anker, Meifod.
G m	*Maclurea,* p. 300. Pl. 1 L, fig. 12.	**Raphistoma**? (or **Helicotoma) macromphala,** McCoy sp.	a. 168, Craig Head, Girvan.
G m	p. 302.	**Loxonema** ———, sp. A large species, much like those from rocks of the same age in Sweden.	a. 177, Cefn-y-coed Lime quarry; Mynydd Fron Frys, Chirk.
G m	p. 303.	**Holopella** ———, sp. These elongate Turritella-like shells are probably members of the Pyramidellidæ, not *Litorinidæ.*	Troutbeck.
G m	Pl. 1 K, fig. 32, p. 304.	**Holopella monilis,** McCoy.	a. 181, Selattyn Road, near Chirk.
		Cephalopoda (*Tetrabranchiata*). The shells allied to *Nautilus* are of the greatest variety and abundance in the slate rocks. Straight, curved, coiled, or involute, they exhibit every change of form which is exhibited in later formations by the Ammonite group. *Orthoceras* and its subgenera are the common forms in the Bala rocks, while *Lituites, Trochoceras,* and *Phragmoceras,* abound in Silurian strata.	
G m	Pl. 1 L, figs. 28, 29, p. 318.	**Orthoceras vagans,** Salter (Mem. Geol. Surv. III. p. 356, pl. 24, figs. 1—5). A common fossil with distant septa. It ranges to Portugal and Sweden.	a. 175, Coniston; Troutbeck; a. 175*, Dufton, in green slates; (Prof. Harkness) Rhiwlas, Bala.
G m		**Orthoceras,** smooth species.	Holbeck gill (in nodules).
G m	Pl. 1 L, fig. 30, p. 316.	**Orthoceras politum,** McCoy (Quart. Journ. Geol. Soc., Vol. 7, pl. 10, figs. 5, 6).	a. 172, Glenquhaple.
G m	do.	**Orthoceras,** smooth species, with closer septa than *O. vagans* above quoted.	Pusgill, Dufton (green slate).
G m	p. 319.	**Orthoceras bilineatum,** Hall, Pal. N. York, t. 43, f. 2, 3 (Siluria, 3rd ed. p. 200).	Coniston (? Upper Bala). a. 172*, Ardwell, Girvan.

Case and Column of Drawers.	Reference to McCoy's Synopsis: and Figures of Genera.	Names and References: Observations, &c.	Numbers and Localities.
G m		**Orthoceras,** sp. allied to *O. dulce,* Barrande. The ribs apparently visible on the cast.	a. 176*, Holbeck gill (nodules in slate rock).
		(*Ringed or Annulate species— Cycloceras,* McCoy.)	
G m	p. 319.	**Orthoceras annulatum,** Sow. A cosmopolite species, at least in N. Hemisphere.	a. 176, Coniston.
G m	do.	**Orthoceras Ibex,** Sow. (Siluria, 2nd ed. pl. 29, figs. 3, 4).	Coniston.
G m	do.	**Orthoceras arcuoliratum,** Hall (Pal. N. Y. I. pl. 42, fig. 7).	a. 171, Wrae Quarry, Upper Tweed.
		Ormoceras, Stokes. Differs from ordinary Orthoceras in having the siphuncle thicker and formed of bead-like pieces.	
G m		**Ormoceras** ———, new sp.	a. 170, Cymmerig Brook, Bala.
G m	p. 323.	**Lituites cornu-arietis,** Sow. (Siluria, 2nd ed. pl. 7, fig. 10).	a. 230, Coniston.
G m	Pl. 1 L, fig. 26, p. 323. Appendix A. p. VIII.	**Lituites anguiformis (Trocholites,** Hall). The Trocholites have usually whorls, wider than they are deep; *i.e.* the transverse diameter is greater than that from back to front.	a. 165, Mynydd Fron Frys, near Chirk.
	Trocholites, p. 324.	**Lituites (Trocholites) planorbiformis,** Conrad (Salter, Mem. Geol. Surv. III. pl. 25, fig. 5).	a. 162, Cymmerig Brook, Bala.

UPPER BALA GROUP, Sedgw. (restricted in 1866. The Upper Bala of the Synopsis includes the Bala limestone, now *Middle Cambrian*).

1. *Hirnant Limestone* and Llanfyllin beds—viz., pale coloured slates above the Bala limestone. Ash Gill slates, &c., above the *Coniston Limestone*.

2. *Llandovery Rocks* (Phillips, Salter, Lyell—*Lower Llandovery* of the Survey).

The fossils of these two divisions are arranged together in the cases and drawers, as it is clearly impossible always to draw a line between them; and they form indeed one series. But the list is kept in *two separate columns here*, as each group contains a few peculiar species. And it may eventually be proved that No. 2 is unconformable on No. 1. The conglomerates and grits of the Llandovery rock do not appear everywhere under the covering of the Silurian rocks, because these are unconformable on them. But wherever we rise to beds far above the level of the Bala limestone, a profusion of corals, Bryozoa and Brachiopoda of the smaller kinds, take the places of the characteristic Bala limestone shells. I arrange *Ash Gill* beds (above the Coniston limestone) with this division; but not that upper portion of the Coniston flags known as the Brathay flagstone, for that is the base (or nearly) of the Silurian series.

Case and Column of Drawers.	Reference to McCoy's Synopsis: and Figures of Genera.	Names and References: Observations, &c.	Numbers and Localities.	
			Upper Bala proper.	*Llandovery Group.*
		Nidulites, Salter. So called from its resemblance to the egg capsules of marine Gasteropods especially of Natica. The differences are obvious.		a.185, Mullock quarry, Dalquorhan, Girvan.
G n	*Sponges?* *Palæopora favosa.* Pl. 1 C, fig. 3, p. 15, in part only. 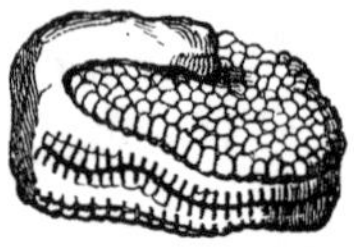	**Nidulites favus,** Salter (Quart. Jour. Geol. Soc. Vol. VII. p. 174; Siluria, 3rd ed. Foss. 30, fig. 3). A curious fossil, having exactly the aspect of miniature honey-comb. The structure is as follows: a thin undulated plate, to which are attached on each side minute hexagonal cups (a pit at the base of each shews the point of attachment), alternating in either face, just like the comb cells, but with a length and breadth of only a line.	[This is one of the commonest of fossils in the same formation at Haverfordwest. Kindred forms are probably to be sought in *Sphærospongia*, a middle Bala fossil.]	

Case and Column of Drawers.	Reference to McCoy's Synopsis: and Figures of Genera.	Names and References; Observations, &c.	Numbers and Localities.	
			Upper Bala proper.	*Llandovery Group.*
	p. 22.	**Nebulipora** (**Monticulipora** of Milne Edwards and Haime is earlier).		
G o		**Nebulipora** ——— sp.		Mathyrafal ffrid.
G o	p. 24.	**Stenopora fibrosa,** Goldfuss (See ante, p. 29). The genus *Stenopora* is not admitted by all writers; but seems to be founded on sufficient characters. According to Lonsdale and McCoy, the tubes have no connecting pores. [N.B. As a rule, individual corals are *extremely* abundant, but the species few, in the Llandovery rocks.]		b. 183, very common at Mathyrafal ffrid, near Meifod. [As this locality is often quoted, we shall abbreviate it to the first word.] Allt Goch, and other places near Llanfyllin.
G o		**S. fibrosa,** var. **ramulosa,** Phill. (Mem. Geol. Surv. II. pt. 1, p. 307). A narrow-branched coral which is quite common, and may be a distinct species, or even a *Chætetes.*		do. do. b. 171, Llanfyllin.
G o	p. 19.	**Favosites alveolaris,** Goldf. (*F. aspera,* D'Orb. Siluria, 2nd ed. pl. 40, fig. 2).		b. 184, Allt Goch, Llanfyllin; Goleugoed, near Llandovery; Mathyrafal, S. Wales; Mullock, Girvan.
G o	*F. alveolaris,* var. p. 19.	**Favosites multipora,** Lonsdale (Siluria, 2nd ed. pl. 40, fig. 5). A very doubtful fossil, since it is all but impossible, save for the expanded form, to distinguish it from the *F. aspera* which accompanies it. Numerous rows of pores are figured on the tubes by Lonsdale.		b. 186, Mathyrafal, as before; b. 185, Allt Goch, Llanfyllin; Goleugoed.
G o	*Palæopora,* p. 18.	**Heliolites tubulata,** Lonsdale, Sil. 4th ed. pl. 39, fig. 3.		b. 187, Allt Goch, Goleugoed; Mullock.

Case and Column of Drawers.	Reference to McCoy's Synopsis: and Figures of Genera.	Names and References; Observations, &c.	Numbers and Localities.	
			Upper Bala proper.	*Llandovery Group.*
G o	*Palæopora,* p. 16. Pl. 1 c, fig. 4.	**Heliolites megastoma,** McCoy (Siluria, 1867, p. 188, Foss. 30, fig. 7).		b. 188, Mathyrafal.
G o	*Palæopora,* p. 17.	**Heliolites subtilis,** McCoy. Narrow branches, very small cells.		a. 200*, Dalquorrhan, Mullock, Girvan.
G o	p. 26.	**Halysites catenulatus,** Linn. (chain-coral). This long-lived and widely spread coral ranges throughout all the Upper Cambrian and the Silurian formations. It is most abundant in limestone, but has no sort of antipathy to slate, shale, sand, or fine grit. It must have inhabited various depths of water.		b. 189, Mathyrafal; b. 190, Goleugoed.
G o	*Cup-Corals.*	**PETRAIA.**—Munster and Phillips. (*Turbinolopsis* of Lonsdale.)		
	See figure, p. 75.	One of the commonest and simplest of all the Cambrian or Silurian cup-corals. The calyx is so deep, and so strongly ribbed by the toothed lamellæ, that the conical matrix left in this part (frequently all we have preserved), shews all the characters necessary for distinguishing the many varieties, or species, as they are supposed to be. The changes from the young to the adult state are not yet sufficiently known, to prevent us from multiplying species on the characters drawn from the lamellæ. The base of the cup, a solid mass of twisted lamellæ, or tabulæ, is short or long, large or minute, in proportion to the cup, in the various species. To Prof. Phillips and Prof. McCoy we are chiefly indebted for descriptions of the species.		

Case and Column of Drawers.	Reference to McCoy's Synopsis: and Figures of Genera.	Names and References; Observations, &c.	Numbers and Localities.	
			Upper Bala proper.	*Llandovery Group.*
case G	Pl. 1 B, fig. 26, p. 40.	**Petraia subduplicata,** McCoy. A very common form, of which the next variety is the more frequent form. The main lamellæ and the intermediate ones are nearly of the same size, at least near the edge of the cup.		a. 201, Dalquorrhan, Girvan (abundant). On the same tablet are two Middle Bala species for comparison (p. 43); a. 201*, Haverfordwest, large size (J. W. S. 1863).
G	Pl. 1 B, fig. 26 *b*, p. 41.	**Petraia subduplicata,** var. **crenulata,** McCoy. This is the common Llandovery variety.		a. 202, equally common with the above at both localities, and wherever the Llandovery rocks are known; Mathyrafal; plentiful at Haverfordwest.
G o	Pl. 1 B, fig. 25, p. 41.	**Petraia uniserialis,** McCoy. Small specimens and probably young.		Penlan, Llandovery. One of the best localities for May Hill Sandstone—in an arenaceous and shaly form.
G o	p. 40.	**Petraia elongata.** (Siluria, 3rd ed. pl. 38, fig. 6.)		a. 202*, Bala; locality doubtful; probably from these beds. J. W. S.
G o	Pl. 1 B, fig. 23, p. 39.	**Petraia æquisulcata,** McCoy. A large species.		a. 203*, Mullock, Girvan; Braes, Girvan.
G o	p. 40.	**Petraia rugosa,** Phillips (Pal. Foss. Corn. & Devon, pl. 2, fig. 7). A species with very wide cup.		Llandovery.
G o	do.	**Petraia,** sp.		Dalquorrhan, Mullock.
G o	do.	**Petraia,** sp.		Mathyrafal, Meifod.
G	Pl. 1 B, fig. 25, p. 41.	**Petraia uniserialis.** See above.	a. 203, Llansantfraid, Glyn Ceiriog;	
G	p. 40.	**Petraia rugosa,** Phill. See above. (Phil. Pal. Foss. pl. 2, fig. 7).	a. 204, Penarth, (same tablet with last, perhaps Middle Bala); Meifod.	
		Omphyma, Milne Edwards. Large simple cup-corals with four deep pits in the base of the cup.		

Case and Column of Drawers.	Reference to McCoy's Synopsis: and Figures of Genera	Names and References; Observations, &c.	Numbers and Localities.	
			Upper Bala proper.	*Llandovery Group.*
G	(*Caninia*, p. 28.)	**Omphyma turbinata,** Linn. (Edw. Monogr. Brit. Fos. Corals, pl. 69, fig. 1). The largest and most common of all the cup-corals, ranging from this formation to the Wenlock, where it is abundant. Siluria, 3rd ed. pl. 39, fig. 11.		b. 193, Mathyrafal, Meifod.
G o	*Bryozoa.*	**Ptilodictya,** ——— sp.	b. 194, Llanfyllin.	
G o		**Glauconome disticha,** Goldf. (See Pet. Germ. t. 65, fig. 15).	b. 195, Llanfyllin.	
G o	*Glauconome*, p. 49. Pl. 1 C, fig. 16, p. 46.	**Ptilodictya explanata,** McCoy (See p. 44).		Mathyrafal, Meifod; Pen-y-Craig, do.
G o	Pl. 1 C, fig. 15, p. 46.	**Ptilodictya costellata,** McCoy (See p. 44).		b. 197, Dalquorrhan, Mullock, Girvan, Ayr.
		ECHINODERMATA.		
G o	do.	**Cyathocrinus,** cup of some species.		
	Worm-tubes.	**ANNELIDA.**		
G o	p. 63.	**Tentaculites ornatus,** Sow. (Siluria, 3rd ed. pl. 16, fig. 11). Apparently the common Dudley species.		b. 199, Mullock, Girvan.
G o	*T. annulatus*, p. 63.	**Tentaculites anglicus,** Salter (Siluria, 3rd ed. pl. 1, fig. 3). *T. annulatus*, Sil. Syst. The *T. annulatus* is a Devonian form without the fine longitudinal striæ.		b. 200, Pwllheli (? Llandovery, which formation must exist there). J. W. S.
		CRUSTACEA — Phyllopoda and Trilobita.		
G o	Pl. 1 E, fig. 3, p. 136.	**Beyrichia complicata,** Salter (Mem. Geol. Surv. III. pl. 19, fig. 9).		b. 201, Mathyrafal.
	Trilobites.	The Trilobites are few in number, and of species chiefly common to Cambrian and Silurian rocks.		
G o	*Portlockia*, p. 163.	**Phacops Stokesii,** Milne Edw. (Salter Mon. Brit. Tril. pl. 2, figs. 1—6). A common species at Dudley.		b. 202, Mullock, Girvan.

Case and Column of Drawers.	Reference to McCoy's Synopsis: and Figures of Genera.	Names and References; Observations, &c.	Numbers and Localities.	
			Upper Bala proper.	*Llandovery Group.*
G o	*Harpidella*, p. 143.	**Cyphaspis megalops,** McCoy (Decade Geol. Surv. No. 7, pl. 5). For genus see Wenlock rocks.		b. 203, Mullock, Girvan.
G	*Ceraurus Williamsi*, p. 155. Pl. 1 F, fig. 13.	**Cheirurus bimucronatus,** Murch. (Decade 7, pl. 2. Mon. Brit. Tril. pl. 5, figs. 1—5). The genus *Ceraurus* cannot stand—so ill described.		a. 199*, Goleugoed (Mr Williams).
G o		**Illænus Thomsoni,** Salter (Pal. Mon. Brit. Tril. pl. 28, figs. 2—4). A species quite characteristic of Llandovery beds. The *Illæni* differ by good but obscure characters.		a. 183, Mullock.
G o		**Lichas bulbiceps?** Salter and Phillips (Mem. Geol. Surv. II. Pt. 1, pl. 8, fig. 8). I am not sure of this being the Dudley one.		b. 204, Mullock, Girvan.
G o	*Dysplanus centrotus*, Pl. 1 E, fig. 19, p. 173.	**Illænus Bowmanni,** Salter (Mon. Brit. Tril. pl. 28, figs. 6—13). One of the characteristic Cambrian forms which range here also.		b. 205, Drummock, Girvan.
G	p. 158.	**Encrinurus punctatus,** Brünn (Quart. Journ. Geol. Soc. Vol. 6, p. 158, pl. 32, figs. 1—5). Another fossil which links the Llandovery to the Silurian.		b. 206, Mathyrafal.
G o	do.	**Encrinurus punctatus,** var. **arenaceus,** Salter (Decade 7, Geol. Surv. pl. 4, p. 6). In some beds this variety prevails.		
G	*C. Blumenbachii*, p. 165. *C. subdiademata*, Pl. 1 F, fig. 9, p. 166.	**Calymene Blumenbachii,** Brongn. (Mon. Brit. Tril. pl. 8, figs. 7—16). The common Dudley form of the species, which varies much. In the Caradoc the glabella (stomach) is reduced, but attains its full size in the Dudley Limestone.		a. 183*, Drummock, near to Girvan; Mullock (*C. subdiademata*).

Case and Column of Drawers.	Reference to McCoy's Synopsis: and Figures of Genera.	Names and References; Observations, &c.	Numbers and Localities.	
			Upper Bala proper.	*Llandovery Group.*
G o	*Odontochile*, p. 161. Pl. 1 G, figs. 15, 16.	**Phacops obtusicaudatus,** Salter (Mon. Brit. Tril. 1864, pl. 1, figs. 42—45). A species with the habit of the common Dudley *P. caudatus* (Dudley butterfly). The tail is blunt.	a. 132, Cold Well, near Wrae, Windermere; Helms Knot, N. E. side?	
G o		**Proetus,** sp. Fragments only.	Helms Knot, Dent.	
	T. gibbifrons, p. 145.	**Trinucleus concentricus.**		a. 137, Pen-y-Craig.
	Terebratulæ, &c. *Lamp shells.*	**MOLLUSCA BRACHIOPODA.** Lamp shells are very common in the Llandovery rocks, and indeed throughout the Bala series. But the genera *Meristella, Atrypa, Pentamerus* are hardly known in Middle Bala, and become of great importance in this division. The other Brachiopods are chiefly Upper Cambrian, but a few are Silurian forms.		
G o	*Spirigerina*, p. 197.	**Atrypa marginalis,** Dalman (Siluria, 3rd ed. pl. 9, fig. 2). One of the long-lived forms, beginning in the Middle Bala rock.	N. of Llanfyllin.	b. 207, Allt Goch, Llanfyllin; Mathyrafal; Goleugoed.
G o	do. p. 198.	**Atrypa reticularis,** Linn. (Siluria, 2nd ed, pl. 9, fig. 1). The earliest appearance of this truly cosmopolite fossil.		b. 208, Goleugoed.
G o	*Hemithyris*, p. 201.	**Atrypa hemisphærica,** Sow. (Siluria, 2nd ed. pl. 9, fig. 3).		a. 199, Mullock.
G o	*Hemithyris*, p. 202. Pl. 1 H, fig. 10.	**Atrypa hemisphærica,** var. **scotica,** McCoy. A large variety. Dav. Sil. Brach. t. 13, fig. 31.		
G o		**Lingula,** sp. Lingulæ are rare in the Llandovery rocks.		b. 209, Pen-y-Craig.

Case and Column of Drawers.	Reference to McCoy's Synopsis: and Figures of Genera.	Names and References; Observations, &c.	Numbers and Localities.	
			Upper Bala proper.	*Llandovery Group.*
G	*Hemithyris*, p. 199. Pl. 1 H, figs. 6—8.	**Meristella angustifrons,** McCoy (*Rhynchonella* of some). *Meristella* has internal *spires*, otherwise it has much the look of *Terebratula*. Davidson, Sil. Brach. t. 10, figs. 21—27.		a. 193, Dalquorrhan, Mullock, Girvan (very common). It also occurs in S. Wales.
G	*Spirifer percrassa*, p. 194.	**Meristella? crassa,** Sow. (*Atrypa*, Sil. Syst. t. 21, f. 1). In the Sil. System are figures only of the interior, which near the beak is much ribbed. Sil. 2nd ed. pl. 9, fig. 6.		a. 194, Cefn Rhyddan; Cyrn-y-Brain; Mathyrafal; b. 24, Cwar Mawr, Cilgroyn (Mr Hughes).
G	*Hemithyris*, p. 207. Pl. 1 H, fig. 9.	**Meristella? subundata,** McCoy. (The genus of this broad form is a little uncertain, says Davidson Sil. Brach. t. 13, fig. 4.)	Altffair ffynnon.	a. 195, Mathyrafal.
G o	do.	**Meristella,** sp. allied to *M. crassa.*		Mandinam, near Llandovery.
G o	do.	**Meristella,** sp. (Looks like part of *Pentamerus lens.*)		Mandinam, Llandovery.
	Pl. 1 H, fig. 3, p. 188.	[**Siphonotreta micula** has been quoted from Upper Bala—it is a Lower Bala fossil only.]		
G	*Hemithyris*, p. 203.	**Rhynchonella? Lewisii,** Davidson (Siluria 3rd ed. p. 226, foss. 58. Davidson, Mon. Sil. Brach. pl. 23, figs. 25—28). *Rhynch. Lewisii* is one of the commonest Dudley shells.		b. 210, Mathyrafal.
G o	*Hemithyris*, p. 205.	**Pentamerus rotundus?** (Siluria, 3rd ed. pl. 22, fig. 18). (*Pentamerus?* Davidson, Sil. Brach. t. 15, figs. 9—11.)		b. 210*, Allt Goch.
G	p. 209.	**Pentamerus lens,** Sow. (Siluria, 3rd ed. pl. 8, figs. 9, 10). The flattened form, and short lamellæ distinguish it.		a. 982, Llettyrhydd-od, Llandovery (Mr Hughes); a. 192, Mandinam; Noeth Grug, Mathyrafal, rare.

Case and Column of Drawers.	Reference to McCoy's Synopsis: and Figures of Genera.	Names and References; Observations, &c.	Numbers and Localities.	
			Upper Bala proper.	*Llandovery Group.*
	p. 211. *P. lævis*, p. 209.	**Pentamerus oblongus,** Sow. (Siluria, 3rd ed. pl. 8, figs. 1—3). *P. lævis* is the young only. The lamella is very long.		Cyrn-y-Brain, near Wrexham; a. 192*, Noeth Grug.
G		**Pentamerus undatus,** Sow. (Siluria, 3rd ed. pl. 8, figs. 5—7). Davidson, Sil. Br. t. 19, figs. 4—9.		b. 209*, Mathyrafal; Goleugoed; Mandinam.
G	*P. microcamerus*, McCoy. p. 210.	**Pentamerus liratus,** Sow. *Stricklandinia lirata.* Davidson, Mon. Sil. Brach. pl. 20.		a. 190, Mandinam.
		Pentamerus globosus, Sow. (Davids. Sil. Brach. t. 19, figs. 10—12, is rare).		
		N.B. The species of *Pentamerus* with short hinge-plates and a wide instead of elongate habit have been long distinguished by Billings as a new genus, *Stricklandinia.* They are only found in the Llandovery and May Hill rocks.		
G o	p. 213.	**Orthis Actoniæ,** Sow. (Siluria, 2nd ed. pl. 5, fig. 11. Mem. Geol. Surv. III. p. 339, pl. 21, figs. 1—8).		b. 208*, Mathyrafal.
G o	p. 192.	**Orthis biforata,** Schlot. var. *lynx*, Eichwald (Geol. Russ. t. 3, fig. 3).	b. 206*, Maes-y-fallen.	b. 207*, Mathyrafal.
G o	*Spirifera biforata*, var. *fissicostata*, McCoy, p. 193.	**Orthis fissicostata,** McCoy. A new species, having a very wide form, and the ribs divided.	Ash Gill, Westmorland; b. 205*, Maes Hir; and b. 204*, Aber Hirnant, Bala.	Mathyrafal.
G o	p. 213.	**Orthis biloba,** Linn. (Siluria, 2nd ed. pl. 9, fig. 20).	b. 203*, Cefn Goch, Glyn Ceiriog.	
G o	p. 214.	**Orthis calligramma,** Dalm. (Geol. Russ. t. 13, fig. 7.)		
G o		**Orthis,** var. **Walsallensis,** Davidson (Salter, Mem. Geol. Surv. III. p. 337, pl. 22, figs. 6, 7).	b. 201*, Llanfyllin, N. Wales.	
G o	*O. plicata*, p. 222.	**Orthis,** var. **plicata,** Sow. (Siluria, 2nd ed. pl. 5, fig. 9).	Rhosfawr, Llanfyllin.	b. 202*, Noeth Grug; Penlan? Llandovery.

Case and Column of Drawers.	Reference to McCoy's Synopsis: and Figures of Genera.	Names and References; Observations, &c.	Numbers and Localities.	
			Upper Bala proper.	*Llandovery Group.*
G o	p. 216.	**Orthis elegantula,** Dalm. (Siluria, 2nd ed. pl. 5, fig. 5).	b. 371, Maes-y-fallen, Bala; b. 372, Rhosfawr, Glog, near Llanfyllin.	b. 373, Mullock Quarry; b. 374, Mathyrafal.
G o	p. 221.	**Orthis parva** (Pander. Geol. Russ. t. 26, fig. 10).	b. 376, Llanfyllin.	b. 377, Penlan, Llandovery.
G o	p. 221.	**Orthis parva,** var. **avellana** (De Vern. Geol. Russ. t. 13, figs. 3, 4).		b. 378, Mullock.
G o	*Spirifera,* p. 194.	**Orthis insularis,** Eichw. (De Vern. Geol. Russ. Vol. II. pl. 8, fig. 7).		Mathyrafal; a. 978, Llettyrhyddod, Llandovery (Mr Hughes).
G o	p. 220.	**Orthis hybrida,** Sow. (Siluria, 2nd ed. pl. 20, fig. 13).		b. 379, Mathyrafal.
G o G	Pl. 1 H, fig. 11, p. 219.	**Orthis Hirnantensis,** McCoy (Ann. Nat. Hist. 2nd Ser. Vol. VIII. p. 395). Very characteristic of the Upper Bala limestone.	b. 381, Cwm-yr-Aethren; b. 380, Maes-y-fallen; a. 184, Maes-Hir; b. 382, Aber Hirnant; b. 383, Cerrig-y-Druidion.	
G o	*O. confinis,* p. 215.	**Orthis,** sp.—Like *Strophomena Pecten.*		b. 384, Mullock.
G o	p. 224.	**Orthis protensa,** Sow. (Siluria, 2nd ed. pl. 9, fig. 22). A squarish shell, flattened, and with the hinge-teeth parallel and short.	b. 385, Ash Gill, Westmorland; Helms Knot, Dent.	a. 191, Goleugoed; Llangynyw; Mathyrafal, &c.
G o	p. 225.	**Orthis reversa,** Salter (Sil. Foss. Ireland, pl. 5, fig. 2).		b. 386, Mullock, Girvan.
G o	Pl. 1 H, figs. 15—19, p. 227.	**Orthis sagittifera,** McCoy (Ann. Nat. Hist. 2nd Ser. Vol. VIII. p. 398). Very near to *O. turgida.*	a. 180, Aber Hirnant; Cwm-yr-Aethren, Bala.	
G o	p. 228.	**Orthis testudinaria,** Dalm. (Siluria, 2nd ed. pl. 5, figs. 1, 2).	b. 388, Maes-y-fallen, Bala.	b. 389, Mullock, Girvan; b. 390, Mathyrafal.
G o	p. 230.	**Orthis vespertilio,** Sow. (Siluria, 2nd ed. pl. 6, figs. 1—3).		b. 391, Mathyrafal.
G o	*O. turgida,* p. 229 Pl. 1 H, figs. 20—24.	**Orthis,** sp. *O. turgida* is a very peculiar species, with good characters in the teeth. This is distinct.		b. 392, Craig Wen; Mathyrafal.

Case and Column of Drawers.	Reference to McCoy's Synopsis: and Figures of Genera.	Names and References; Observations, &c.	Numbers and Localities.	
			Upper Bala proper.	*Llandovery Group.*
	Leptagonia, p. 248.			
G o	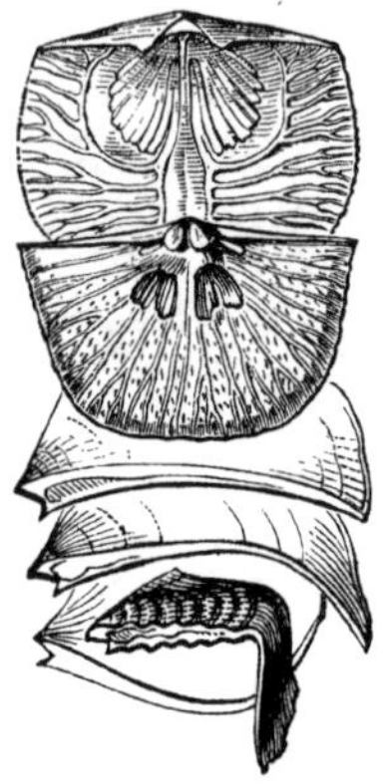	**Strophomena depressa,** Sow. (*S. rhomboidalis* of Wahlenberg. Davidson, Introd. to Class. Brach. pl. 8, figs. 167—173). This weed of the Silurian and Cambrian rocks assumes fifty different shapes.	b. 400, Rhosfawr, N. of Glog, Llanfyllin.	b. 401, Allt Goch, Llanfyllin; b. 402, Mathyrafal; Penlan.
G o	*Leptæna*, p. 241.	**Strophomena antiquata.** Sow. (Siluria, 2nd ed. pl. 20, fig. 18).		Mathyrafal; a. 977, Lettyrhyddod, Llandovery (Mr Hughes).
G o	*L. alternata*, p. 233.	**Strophomena,** sp. Fine lines. So many are the species of this group in all parts of the world, that good internal characters are wanted to discriminate them.		b. 404, Dalquorrhan, Girvan.
G o	p. 237.	**Leptæna sericea,** Sow. (Siluria, 2nd ed, pl. 5, fig. 14).	b. 405, S. of Llanfyllin.	b. 406, Mathyrafal; b. 407, Goleugoed; b. 408, Allt Goch.
G o	p. 239. 	**Leptæna sericea,** var. **rhombica,** Phill.		b. 409, Mathyrafal.
G o	*L. quinquecostata*, Pl. 1 H, f. 30, 31, p. 236 (not fig. 32).	**Leptæna transversalis,** var. **undulata,** Salter (not the true *quinquecostata*, for which see p. 64).		b. 410, Craig Wen; b. 411, Cefn Rhyddan. a. 198, Mathyrafal, near Meifod. a. 196, Pen-y-Craig.
G o		**Leptæna scissa,** Salter, MS. A small and highly convex form, with the muscles deeply indented. (Siluria, 3rd ed. p. 210.)		a. 197, Pen-y-Craig.
	Bivalves,	**LAMELLIBRANCHIATA.**		
G o	*Orthonotus semisulcatus*,	**Orthonotus** ——— sp.		a. 189, Mullock.
G o	p. 275.	**Ctenodonta** (*Nucula* formerly), see p. 66. *C. Hughesii*, n. sp. Salter. Squarer and less ovate than the kindred Caradoc forms.		a. 979, Lettyrhyddod, Llandovery (Mr Hughes).

Case and Column of Drawers.	Reference to McCoy's Synopsis: and Figures of Genera.	Names and References; Observations, &c.	Numbers and Localities.	
	Univalves.		*Upper Bala proper.*	*Llandovery Group.*
		GASTEROPODA.		
G o		**Pleurotomaria,** large ribbed.		Penlan.
G o		**Pleurotomaria,** large ribbed.		b. 412, Mandinam.
G o	Pl. 1 L, fig. 20, p. 292.	**Murchisonia cancellatula,** McCoy.		a. 182, Mullock, Dalquorrhan.
G o	Pl. 1 K, fig. 42, p. 294.	**Murchisonia pulchra,** McCoy. Sil. Foss. Ireland, t. 1, fig. 19.		b. 414, Mullock; b. 413, Mathyrafal.
G o	*M. simplex,* p. 294, Pl. 1 K, fig. 44.	**Murchisonia,** sp. ——— Broader keel and longer base than *M. simplex.*		a. 207, Mullock.
G o	*Euomphalus tricinctus,* p. 299.	**Murchisonia tricincta,** McCoy. Hardly a determinable species, but certainly a *Murchisonia.*		a. 179, Dol Fan, Rhayadr.
G o	*Trochus Helicites,* p. 297.	**Raphistoma æqualis?** Salter (Siluria, 3rd ed p. 197, Foss. 40, fig. 2).		b. 415, Dalquorrhan.
G o	As *E. tricinctus,* p. 299.	**Euomphalus,** ——— sp. (Indistinct.)		b. 416, Dalquorrhan.
G o	Pl. 1 L, fig. 18, p. 297.	**Trochus? Moorei,** McCoy.		a. 187, Dalquorrhan.
G o	Pl. 1 L, fig. 17, p. 304.	**Holopella tenuicincta,** McCoy.		a. 188, Mullock.
G o	(Not of 1 L, fig. 15) p. 301.	**Eccyliomphalus (scoticus?** McCoy). This has narrower and rounder whorls, and is not likely to be the species in the Knockdolian beds.		a. 178, Mullock, Dalquorrhan.
		HETEROPODA.		
G o	Pl. 1 L, fig. 25, p. 311.	**Bellerophon subdecussatus,** McCoy.		a. 186, Mullock.
G o		**Bellerophon dilatatus,** Sow.		b. 417, Mullock.
		CEPHALOPODA.		
G o	p. 317.	**Orthoceras tenuistriatum,** Münster, Beit. 3, t. 19, fig. 4.		a. 186*, Mullock.
G o		**Orthoceras,** sp.		b. 418, Mullock.
G o		**Cyrtoceras?** sp. Coarse wide ridges, possibly a large wide *Orthoceras.*		b. 418*, Mullock.

SILURIAN (ROCKS) SYSTEM, Sedgwick; UPPER SILURIAN, Murchison.

Above all the Cambrian Rocks—and this throughout the Northern Hemisphere, so far as known—the Silurian rocks, consisting of the May Hill (or Clinton) Group, Lower Wenlock or Woolhope Group (Sedgwick and Salter), Wenlock Group (Murchison), Ludlow Group (Murchison) including Downton sandstone, lie unconformably on the Cambrian rocks. The percentage of common fossils between the two systems is very small—not more than 6 or 7 per cent.

May Hill Sandstone (Sedgw.). *Upper Llandovery* (Murch.) *Upper Caradoc,* olim. *Clinton group* of North America. *Upper Anticosti group,* Canada.

Case and Column of Drawers.	Reference to McCOY'S Synopsis: and Figures of Genera.	Names and References; Observations, &c.	Numbers and Localities.
		GRAPTOLITIDÆ.	
G a	*G. ludensis,* p. 4. 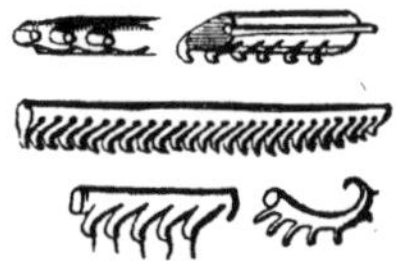	**Graptolithus priodon** (Siluria, 2nd ed. pl. 12, figs. 1, 2).	May Hill and Tortworth.
		AMORPHOZOA.	
	p. 12.	**Stromatopora striatella,** D'Orb. (Siluria, 2nd ed. pl. 41, fig. 31).	May Hill, Gloucestershire.
G a		(**Vioa prisca** is an annelide, see next page.)	
		ZOOPHYTA.	
G a	*Corals.* *Palæopora,* p. 15.	**Heliolites interstinctus,** Linn. (Siluria, 2nd ed. pl. 39, fig. 2).	b. 422, May Hill.
	Pl. 1 c, fig. 4, p. 16.	**Heliolites megastoma** (Siluria, 3rd ed. p. 188, Foss. 30, fig. 7).	Malverns.
G a 1	p. 19.	**Favosites alveolaris,** Goldf. (Pet. Germ. t. 26, fig. 1).	b. 424, May Hill; b. 425, Norbury.
G a	*F. alveolaris,* p. 19, var. a.	**Favosites multipora,** Lonsd. (Siluria, 2nd ed. pl. 40, fig. 5).	May Hill.
		Favosites cristata, Blum. (Siluria, 3rd ed. Foss. 18, fig. 1).	Malverns.
	p. 20.	**Favosites aspera,** D'Orb. (Siluria, 2nd ed. pl. 40, fig. 1).	Malverns.
G a 1	p. 26.	**Halysites catenulatus** (Siluria, 2nd ed. pl. 40, fig. 14).	b. 429, May Hill; Llandovery.
		Palæocyclus, Milne Edw. [See Wenlock limestone.]	
G a 1		**Palæocyclus præacutus,** Lonsd. (Siluria, 2nd ed. pl. 41, figs. 4, 5).	b. 432, May Hill; Worcester Beacon, W. side.

Case and Column of Drawers.	Reference to McCoy's Synopsis: and Figures of Genera.	Names and References; Observations, &c.	Numbers and Localities.
		**Petraia,** Münster.	
G a	p. 40.	**Petraia rugosa,** Phill. Palæozoic Foss. Devon. p. 7, t. 2, fig. 7 c.	b. 435, Castell Craig, Gwyddon.
G a 1	do.	**Petraia elongata,** Phill. (Siluria, 2nd ed. pl. 38, fig. 6).	b. 436, Norbury; Castell Craig, Gwyddon; b. 438, May Hill; Presteign?
G a 1	Pl. 1 B, fig. 26, p. 40.	**Petraia subduplicata,** Siluria, 3rd ed. Foss. 15, fig. 11.	b. 439, Presteign; b. 440, May Hill.
G a 1	Pl. 1 B, fig. 25, p. 41.	**Petraia uniserialis,** McCoy. A young form of some species.	b. 441, Castell Craig, Gwyddon; b. 442, Penlan, Llandovery.
		Petraia bina, Lonsd. Siluria, 3rd ed. p. 219.	a. 239, Malverns or Norbury (quoted from Horderley); b. 443, Pentamerus Shales, Onny River, Shropshire; b. 444, Malverns.
	Caninia, p. 28. 	**Omphyma turbinata,** Linn. sp. (Siluria 3rd ed. pl. 39, fig. 11, M. Edw. Brit. Foss. Cor. pl. 69, fig. 1).	
		POLYZOA (Bryozoa, Ehrenberg).	
		Ptilodictya, Lonsdale. See Wenlock for genus.	
	Lace Corals, &c. p. 45.	**Ptilodictya scalpellum,** Lonsd. (Sil. 3rd ed. pl. 41, fig. 25, Foss. 51, p. 217).	
		ECHINODERMATA.	
G a 1	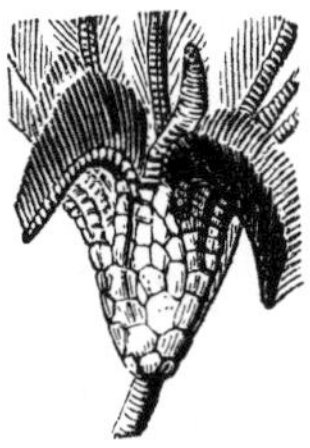	**Glyptocrinus,** sp.	b. 451, Norbury.
	Genus, p. 56.	**Periechocrinus,** sp. Very common in various localities.	
		ANNELIDA.	
		For the following, see remarks on genera in Wenlock list.	
G a 1	p. 63.	**Cornulites serpularius,** Schl. (Sil. 2nd ed. pl. 10, fig. 2, pl. 16, figs. 3—10).	b. 453, Presteign.
G a 1	*T. annulatus,* p. 63. 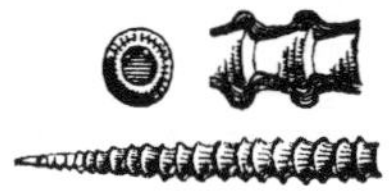	**Tentaculites anglicus,** Salt. (Sil. 2nd ed. pl. 1, fig. 3, & pl. 10, fig. 3).	b. 454, Presteign.
G a 1	Pl. 1 B, fig. 1, p. 260, with *Pterinea demissa.*	**Vioa prisca** is a boring Annelide. The cavities made by Vioa in Oyster-shells are generally filled with sponge.	a. 345, Malverns.

Case and Column of Drawers.	Reference to McCoy's Synopsis: and Figures of Genera.	Names and References; Observations, &c.	Numbers and Localities.
		CRUSTACEA Trilobita.	
G a 1	p. 158.	**Encrinurus punctatus,** Brünn. (Sil. 2nd ed. pl. 10, fig. 5).	b. 455, Norbury and May Hill; a. 344, Pwllheli; Pentamerus Shales, Onny River, Shropshire.
	Ceraurus Williamsi, p. 155.	**Cheirurus bimucronatus,** Mur. (Sil. 2nd ed. pl. 3, fig. 5, pl. 19, figs. 10, 11).	
G a 1	*Portlockia,* p. 163.	**Phacops Stokesii,** Milne Edw. (Siluria, 2nd ed. pl. 10, fig. 6).	b. 458, Norbury.
		Phacops Weaveri, Salt. (Decade Geol. Surv. 2, p. 7, pl. 1. Mon. Brit. Tril. pl. 3, figs. 1—3, 1864).	Tortworth.
	Odontochile, p. 160.	**Phacops caudatus,** Brongn. (Mon. Brit. Tril. pl. 3, figs. 4—18. Siluria, 2nd ed. pl. 17, fig. 2).	May Hill.
		Illænus Thomsoni, Salter (Mon. Brit. Tril. pl. 28, figs. 2—4, pl. 30, figs. 8—10).	Common in the shales of Onny River, Shropshire.
	Forbesia, p. 174.	**Proetus Stokesii,** Murch. (Siluria, 2nd ed. pl. 17, fig. 7).	
G a 1	*Isotelus,* or *Asaphus radiatus?*	**Proetus,** sp. There are many undescribed species of this genus.	a. 346, Braes, 1½ m. E. of Girvan.
		MOLLUSCA Brachiopoda.	
G a 2	p. 250.	**Lingula,** sp.	b. 464, Eastnor, Malvern.
		Lingula parallela, Phill. (Mem. Geol. Surv. Vol. II. pt. 1, pl. 26, fig. 1. Dav. Sil. Brach. pl. 2, figs. 24—27).	Malverns.
G a 4	do. See Wenlock.	**Obolus transversus?** Dav. Sil. Brach. t. 5, figs. 1—6.	b. 467, May Hill.
G a 2 G a	*Spirigerina,* p. 198.	**Atrypa reticularis,** Linn. sp. (Siluria, 2nd ed. pl. 9, fig. 1).	a. 348, Norbury; b. 468, May Hill. b. 469, Pentamerus Shales, Onny River, Shropshire.
	p. 197.	**Atrypa marginalis,** Dalm. (Sil. 2nd ed. pl. 9, fig. 2. Dav. Sil. Brach. t. 13, figs. 23—30).	May Hill; the Braes, near Girvan.
G a 2	Quoted also as *Orthis turgida,* *Hemithyris,* p. 201.	**Atrypa hemisphærica,** Sow. (Siluria, 2nd ed. pl. 9, fig. 3).	b. 472, Norbury; b. 473, Presteign; a. 233, Pwllheli; b. 474, the Braes, Girvan, Ayr.

Case and Column of Drawers.	Reference to McCoy's Synopsis: and Figures of Genera.	Names and References; Observations, &c.	Numbers and Localities.
G a 2		**Rhynchonella decemplicata,** Sow. (Siluria, 2nd ed. pl. 9, fig. 15).	a. 347*, Eastnor Park; a. 347, Pwllheli; b. 476, Presteign.
	Hemithyris lacunosa, p. 202. 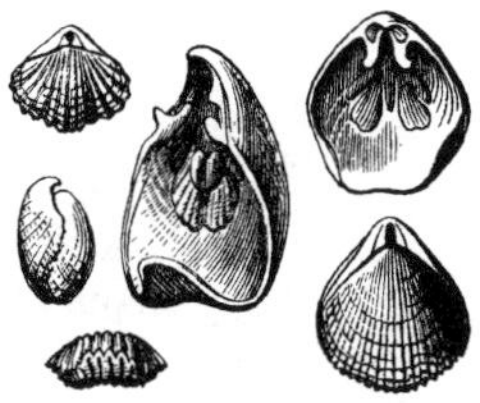	**Rhynchonella borealis,** Schl. (Siluria, 2nd ed. pl. 22, fig. 4, var. *diodonta,* fig. 5).	Common in May Hill Sandstone.
	do.	**Rhynchonella,** sp.	b. 477, Eastnor, Malvern.
G a 2	*Hemithyris,* p. 204.	**Rhynchonella nucula,** Sow.? (Siluria, 2nd ed. pl. 9, figs. 9 & 11).	b. 478, Braes, N. of Girvan.
		Pentamerus, Sow. See Wenlock limestone.	
	p. 209.	**Pentamerus lens,** Sow. (Stricklandinia lens, Davids. Sil. Brach. 19, figs. 13—21).	May Hill and everywhere; Norbury, &c.
G a 3		**Pentamerus linguifer,** Sow. (Silur. 2nd ed. pl. 22, fig. 21. Davids. Sil. Brach. pl. 17, figs. 11—14).	b. 481, Worcester Beacon.
G a 3 and case	*P. microcamerus,* p. 210. 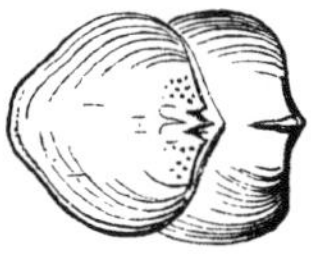	**Pentamerus** (*Stricklandinia*) **liratus,** Sow. (Dav. Sil. Brach. t. 20).	b. 482, Worcester Beacon b. 483, Norbury; a. 234 May Hill; a. 350, as *P microcamerus.*
G a 3 **G a 10**	*P. lævis,* p. 209. 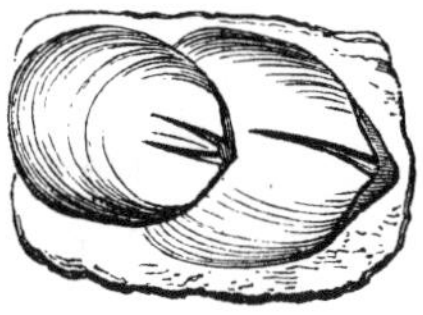	**Pentamerus oblongus,** Sow. (Siluria, 2nd ed. pl. 8, figs. 1—4. Davids. l. c. t. 18). (*P. lævis* is the young only.)	b. 23, Castell Craig, Gwyddon (Mr Hughes); b. 484 Norbury; b. 485, Presteign b. 486, Acton Scott; b. 487 The Hollies; a. 232, Pwllheli.
	p. 208.	**Pentamerus globosus,** Sow. (Siluria, 2nd ed. pl. 8, fig. 8).	The Hollies farm; Pwllheli. [N.B. Pwllheli is unnoticed in any map as May Hill Sandstone.]
G a 2	p. 211.	**Pentamerus undatus,** Sow. sp. (Siluria, 3rd ed. pl. 8, figs. 5—7).	b. 488, Penlan.
G a	*Hemithyris upsilon,* p. 207. 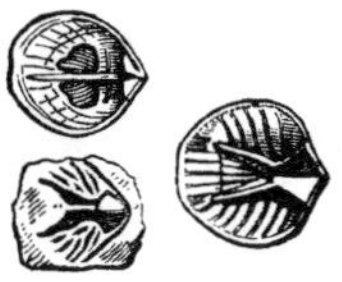	**Meristella didyma,** Dalm. (Siluria, 2nd ed. pl. 22, fig. 15).	Norbury; a. 453, Pwllheli.
G a 2		**Meristella,** sp. Very imperfect.	b. 490, May Hill.
G a 2	*Athyris,* p. 196.	**Meristella tumida,** Dalm. (Siluria, 2nd ed. pl. 22, fig. 20).	b. 491, May Hill.

Case and Column of Drawers.	Reference to McCoy's Synopsis: and Figures of Genera.	Names and References; Observations, &c.	Numbers and Localities.
		Meristella furcata, Sow. (Davids. Sil. Brach. pl. 13, fig. 7—9).	Norbury and Malvern.
G a 4	*Leptagonia,* p. 248.	**Strophomena depressa,** Sow. (Silur. 2nd ed. pl. 20, fig. 20).	b. 494, May Hill.
G a 4		**Strophomena arenacea,** Salter, MSS. (Siluria, 2nd ed. p. 231, 3rd ed. p. 210).	b. 495, Norbury; b. 496, May Hill; b. 497, Presteign.
G a 4	p. 245.	**Strophomena pecten,** Linn. (Silur. 3rd ed. p. 227, Foss. 59, fig. 3).	b. 498, Norbury and May Hill; b. 449, Bromsgrove Lickey.
G a	p. 241.	**Strophomena antiquata,** Sow. (Silur. 2nd ed. pl. 20, fig. 18).	b. 500, Pentamerus Shales, Onny River.
G a	p. 244. as *S. Grandis,* and *Lept. alternata,* p. 233.	**Strophomena compressa,** Sow. (Silur. 2nd ed. pl. 9, fig. 16).	a. 349 May Hill, as *S. grandis.*
G a 4	p. 234.	**Strophomena deltoidea?** (Conr. Pal. N. York, p. 106).	b. 501, May Hill.
	p. 244.	**Strophomena funiculata,** McCoy (Siluria, 3rd ed. Foss. 59, figs. 4, 5).	Very common at May Hill.
		Strophomena applanata, Salter (Mem. Geol. Surv. Vol. II. pt. I. p. 380, pl. 27, figs. 1, 2).	Common at May Hill, &c.
G a	*Spirifera cyrtæna,* p. 193.	**Spirifer plicatellus,** var. **globosus** (Davids. Mon. Sil. Brach. pl. 9, figs. 7, 8).	b. 505, May Hill.
G a 2		**Spirifer plicatellus,** Linn. (ib. pl. 9, figs. 9—12).	b. 506, Pentamerus Shales, Onny River.
G a 2		**Spirifer plicatellus,** var.	b. 507, Norbury.
	Spirifera trapezoidalis, p. 196.	**Spirifer exporrectus,** Wahl. (Siluria, 3rd ed. pl. 21, fig. 3). *Spirifer trapezoidalis,* Dalm. (Silur. 2nd ed. pl. 9, fig. 24. *S. exporrectus,* Wahl. Silur. 3rd ed. pl. 9, fig. 24).	b. 508, Pentamerus Shales, Onny River; b. 509, May Hill.
G a 2		**Spirifer crispus,** Hisinger (Silur. 2nd ed. pl. 21, fig. 4).	b. 510, May Hill.
	S. subspuria, p. 195.	**Spirifer elevatus,** Dalm. (Silur. 2nd ed. pl. 21, figs. 5, 6).	May Hill?
G a 2	p. 214.	**Orthis calligramma,** Dalm. var. **Davidsoni,** Vern. (Silur. 3rd ed. p. 227, Foss. 59, fig. 2).	b. 512, Norbury; May Hill.
	p. 225.	**Orthis reversa,** Salter (Siluria, 3rd ed. p. 210, Foss. 49, fig. 3).	
	Spirifera, p. 192.	**Orthis biforata,** Schloth. (Siluria, 3rd ed. Foss. 36, fig. 4).	Frequent.

Case and Column of Drawers.	Reference to McCoy's Synopsis: and Figures of Genera.	Names and References; Observations, &c.	Numbers and Localities.
G a 2	p. 216.	**Orthis elegantula,** Dalm. (Silur. 2nd ed. pl. 5, fig. 5, pl. 9, fig. 19).	b. 514, May Hill.
	Pl. 1 H, figs. 41, 42, p. 223.	**Orthis porcata,** McCoy (Silur. 3rd ed. Foss. 36, fig. 5, p. 193).	Rare.
G a 2	p. 240.	**Leptæna transversalis,** Dalm. (Silur. 2nd ed. pl. 9, fig. 17).	b. 516, Pentamerus Shales, Onny River.
	p. 237.	**Leptæna sericea,** Sow. (Siluria, 2nd ed. pl. 5, fig. 14, pl. 9, fig. 18).	Two or three varieties at Malvern.
		Leptæna scissa, Salter MSS. (Siluria, 3rd ed. p. 210).	Not rare.
	p. 249.	**Chonetes lata,** Von Buch. (Siluria, 2nd ed. pl. 20, fig. 8).	do.
		LAMELLIBRANCHIATA.	
G a 4	*P. demissa,* Pl. 1 I, fig. 7, p. 260.	**Pterinea retroflexa,** Wahl. sp. var. **demissa,** Conrad.	a. 236*, Eastnor Park.
G a	p. 262.	**Pterinea retroflexa** proper is also common (Siluria, 2nd ed. pl. 9, fig. 26).	a. 236, Malvern.
G a 4		**Ctenodonta lingualis,** Phill. sp. (Mem. Geol. Surv. Vol. II. pt. 1, p. 367, pl. 22, fig. 6).	b. 521, Eastnor Park, Malvern.
	Arca, Pl. 1 K, fig. 1, p. 283.	**Ctenodonta subæqualis,** McCoy, sp. **C. Eastnori,** Sow. sp. **C. deltoidea,** Phill. sp.	Malvern.
		Mytilus mytilimeris, Conrad (Mem. Geol. Surv. II. pt. 1, p. 364, pl. 20, figs. 7—9. Silur. 3rd ed. p. 229. Foss. 61, fig. 6).	May Hill.
		Lyrodesma cuneata, Phill. sp. (Mem. Geol. Surv. II. pt. 1, p. 366, pl. 21, figs. 1—4).	Marloes Bay.
	Leptodomus, p. 278.	**Orthonota amygdalina,** Sow. sp. (Siluria, 2nd ed. pl. 23, fig. 6).	May Hill.
	Capulus, p. 290.	**GASTEROPODA.**	
		Acroculia haliotis, Sow. sp. (Siluria, 2nd ed. pl. 24, fig. 9).	

Case and Column of Drawers.	Reference to McCoy's Synopsis: and Figures of Genera.	Names and References; Observations, &c.	Numbers and Localities.
G a	p. 298.	**Euomphalus funatus,** Sow. (Silur. 2nd ed. pl. 25, fig. 3).	b. 526, May Hill.
	p. 299.	**Euomphalus sculptus,** Sow. (Silur. 2nd ed. pl. 9, fig. 27, pl. 25, fig. 2).	May Hill.
G a 4	*Turbo,* Pl. 1 K, fig. 36, p. 295. Pl. 1 L, fig. 22.	**Cyclonema crebristria,** McCoy (Mem. Geol. Surv. III. p. 347, woodcut 14, fig. 5).	b. 528, Norbury.
G a 4		**Cyclonema,** sp.	b. 529, Norbury.
G a 4	*Naticopsis,* p. 302.	**Natica** or **Acroculia.**	b. 530, May Hill.
G a 4		**Holopea** ——— sp. (For genus see Mem. Geol. Surv. III. p. 347, woodcut 14).	b. 531, Castell Craig Gwyddon.
		Holopea tritorquata, McCoy sp. (Sil. Foss. Irel. p. 12, pl. 1, fig. 8).	Not unfrequent.
G a 4	p. 303.	**Holopella gregaria,** Sow. sp. (Siluria, 2nd ed. pl. 34, fig. 10 A).	b. 533, Eastnor Park.
G a 4	do.	**Holopella cancellata,** Sow. sp. (Siluria, 2nd ed. pl. 10, fig. 14).	b. 534, Presteign.
G a 4		**Murchisonia** ——— sp.	b. 535, Eastnor Park.
		PTEROPODA and HETEROPODA.	
	pp. 308, 309, 311.	**Bellerophon bilobatus, trilobatus, dilatatus, expansus,** Sow.	
		Eccyliomphalus ? lævis, Sow. (Silur. 2nd ed. pl. 25, fig. 9).	
		CEPHALOPODA.	
G a 4	p. 322.	**Phragmoceras ventricosum,** Sow. (Siluria, 3rd ed. pl. 32).	a. 235, May Hill.
G a 4		**Orthoceras** ——— sp.	May Hill; b. 538, Eastnor Park.
G a 4	p. 319.	**Orthoceras annulatum,** Sow. Siluria, 3rd ed. pl. 26, fig. 1.	b. 539, May Hill.

LOWER WENLOCK ROCKS (Woolhope Limestone and Shale; Denbigh Grit and Flagstones; Coniston Flags (upper) and Coniston Grits).

The base of the Wenlock series is in Shropshire and the border counties represented by a limestone (the Woolhope), which differs a good deal in its organic contents from the Wenlock limestone.

In N. Wales the Denbighshire grits occupy this place, and contain Wenlock fossils. They pass eastward into the lower portions of the Denbigh flagstones.

In Westmorland the Coniston or Brathay flag (upper part) has several of the fossils of the Denbigh flagstone. Both in it, and especially over it, lie thick beds of grit, the chief mass of which (Coniston grits) has fossils like those of the Denbigh flagstone and Coniston flagstone. We therefore class all these together as Lower Wenlock. But as it is not quite certain how far the two formations (Denbigh grit and Denbigh flagstone, and Coniston grit and flagstone) are to be considered identical, I give the lists side by side, not mixed together.

Lastly, there is good reason to believe the flagstones of Balmae, Kircudbright, are of this age. They contain Wenlock fossils, with some *Graptolites* and others, which render this the more likely position for them.

Denbighshire grit D. G.; Denbighshire flagstone D. F.
Coniston grit C. G.; Coniston flagstone C. F.
Woolhope limestone and shale W. L.
Kircudbright flagstones (Balmae shore) are certainly the equivalent of the Denbigh flags (J. W. Salter).

Case and Column of Drawers.	Reference to McCoy's Synopsis: and Figures of Genera.	Names and References; Observations, &c.	Numbers and Localities.	
		HYDROZOA or BRYOZOA.	*Wales and Shropshire.*	*Westmorland and S. of Scotland.*
G a 5 **G a 6** **G b**	*G. ludensis*, p. 4.	**Graptolites priodon,** Bronn (Siluria, 2nd ed. pl. 12, figs. 1, 2).	D. G. Moel Seisiog; b. 548, Gwyddelwern, Llanrwst. D. F. b. 549, Pen-y-Craig, Llangynyw; b. 550, Llansantfraid, Glyn Ceiriog (as *Diplograpsus pristis*); D. G. b. 551, Nant-Gwrhwyd Uchaf; under Bron Einion.	C. G. Casterton Fell, N. end. C. F. Common at Helmside, &c.; b. 555, Helms Knot; b. 556, Brathay; a. 244, Rebecca Hill, near Ulverston; Outrake, near Ulverston; a. 244*, D. F. Balmae shore, Kircudbright.
G a 5	*G. sagittarius*, p. 5.	**Graptolites Flemingii,** Salter (Quart. Journ. Geol. Soc. VIII. p. 390). Differs in the broader shaft, and narrower straight teeth (cells) from the common *G. priodon*. It is probably a very common species in Lower Wenlock. (*G. sagittarius* is a Lower Bala species.)	b. 554, D. F. Craig ddu Allt, Llangollen.	a. 245*, D. F. Balmae shore, Kircudbright.

Case and Column of Drawers.	Reference to McCoy's Synopsis: and Figures of Genera.	Names and References; Observations, &c.	Numbers and Localities.	
			Wales and Shropshire.	*Westmorland and S. of Scotland.*
G a 5	(*Diplograpsus ramosus*, p. 8.)	**Graptolites** sp. Very imperfect, but certainly *Graptolithus.*		b. 558, D.F. Old Slate Quarry, Ulverston to Ireleth.
G a		**Retiolites Geinitzianus,** Barr. (Siluria, 3rd ed. p. 541. Foss. 90, fig. 2).	a. 454, D. G. Pentre cwm dda.	
		ZOOPHYTA—ANTHOZOA.		
G a 9	*Corals.* *Strephodes pseudoceratites,* Pl. 1 B, fig. 20, p. 30.	**Cyathophyllum pseudoceratites**? McCoy's figured specimen.	a. 459, W. L. Old Radnor road, Presteign.	
G a 8	p. 24.	**Stenopora fibrosa,** Goldf. (Siluria, 3rd ed. pl. 40, figs. 6, 7).	b. 560, W. L. Old Radnor.	b. 561, C. F. Helms Knot; Howgill Fell.
G a 6		**Petraia,** ——— small sp. (See May Hill Sandstone).	b. 562, D. F. Llangynyw Rectory, Montgomeryshire.	
G a 9	p. 20.	**Favosites Gothlandica,** Linn. (Siluria, 2nd ed. foss. 17, figs. 2, 3).	b. 563, W. L. Old Radnor to Presteign.	
G a 6		**Favosites,** sp. (*aspera?*) just possibly a *Dictyocaris* (MS.).		a. 463, D. F. Balmae shore, Kircudbright.
		ECHINODERMATA.		
G a 7	*Echinodermata,* Pl. 1 D, fig. 3, p. 55.	**Actinocrinus pulcher,** Salter. The common flagstone species. The stems are tubercular, the pelvis small. The arms sinuous and much branched. It appears to be often attached to *Orthoceras,* as if it had floated.	b. 565, D. F. Nant-Gwrhwyd Uchaf, Llangollen.	b. 566 (cast). C. F. Osmotherly Common, Ulverston.
G a	*Eucalyptocrinus,* Pl. 1 D, fig. 2, p. 58?	**Hypanthocrinus**? (*H. polydactylus,* See Wenlock).	b. 567, W. L. Old Radnor.	
G a		**Glyptocrinus,** sp. For genus see Wenlock Limestone.	b. 569, D. G. Plas Madoc.	

Case and Column of Drawers.	Reference to McCoy's Synopsis: and Figures of Genera.	Names and References; Observations, &c.	Numbers and Localities.	
			Wales and Shropshire.	*Westmorland and S. of Scotland.*
		ANNELIDA.		
	Worm tubes.	Hard shelly tubes of sea-worms.		
Ga 9		**Cornulites serpularius,** Schl. (Siluria, 2nd ed. pl. 10, fig. 2, pl. 16, figs. 3—10).	b. 570, W. L. Littlehope, Woolhope.	
Ga 6	p. 131.	**Spirorbis Lewisii,** Sow. (Siluria, 2nd ed. pl. 16, fig. 2), like the little sessile Serpulæ on seaweeds.		b. 571, C. F. Casterton Fell; Helmside.
		CRUSTACEA Trilobita.		
Ga	*Trilobites.*	**Homalonotus cylindricus,** Salter, Mon. Brit. Tril. p. 116, fig. 27, pl. 11, fig. 12.	a. 237 (figured specimen). Littlehope, Woolhope.	
	Odontochile caudata, var. *minor,* p. 161.	**Phacops Downingiæ** (Siluria, 2nd ed. pl. 18, fig. 5).		
Ga		**Phacops** ib. var. **cuneatus,** Salter (Mon. Brit. Tril. 1864, p. 28, fig. 8).	a. 458, D. G. Moel Seisiog, Llanrwst.	
Ga 9	*Odontochile,* p. 160.	**Phacops caudatus** var. **corrugatus,** Salter, ined.	a. 461, W. L. Littlehope.	
Ga 8	*Portlockia,* p. 163.	**Phacops Stokesii,** Milne Edwards (Siluria, 2nd ed. pl. 10, fig. 6).		b. 572, C. F. Half a mile N. of Ulverston.
Ga 6		**Acidaspis Hughesii,** n. s., Salter. A new species.		b. 573, C. G. Casterton Low Fell, Kirkby Lonsdale.
		PHYLLOPODA.		
Ga 6		**Peltocaris anatina,** n. s., Salter. A bivalve phyllopod, more characteristic of Cambrian rocks. The semioval rostrum is seldom found.		b. 574, C. F. Rebecca Hill, Ulverston.
	Brachiopods.	**MOLLUSCA BRACHIOPODA.**		
Ga 8		**Discina Forbesii,** Davids. (Siluria, 2nd ed. foss. 57, fig. 11. *Orbiculoidea,* ibid. 3rd ed. p. 226.)		b. 576, C. G. Middleton Fell, Kirkby Lonsdale.

Case and Column of Drawers.	Reference to McCoy's Synopsis: and Figures of Genera.	Names and References; Observations, &c.	Numbers and Localities.	
			Wales and Shropshire.	*Westmorland and S. of Scotland.*
Ga 9	*Spirigerina*, p. 197.	**Atrypa marginalis,** Dalm. (Dav. Sil. Brach. t. 15, figs. 1, 2).	b. 590, W. L. Worcester Beacon.	
Ga 9	p. 198.	**Atrypa reticularis,** Linn. (Dav. Sil. Brach. t. 14).	b. 591, W. L. Littlehope, Woolhope.	
Ga 6	*Spirifera cyrtæna*, p. 193.	**Spirifer cyrtæna,** var. **plicatellus,** Linn.? (Siluria, 2nd ed. pl. 9, fig. 25).	b. 581, D. G. Plas Madoc, N. of Llanrwst.	
Ga		**Athyris** (like *A. depressa*, Siluria, 2nd ed. pl. 22, fig. 17).	b. 586, D. G. N. of Llanrwst.	
Ga 8	*Hemithyris*, p. 201.	**Meristella didyma,** Dalm. (Siluria, 2nd ed. pl. 22, fig. 15).		b. 578, C. G. Middleton.
Ga 9		**Meristella,** a small species.	b. 579, W. L. Old Radnor road.	
Ga	*Spirifera percrassa*, p. 194.	**Meristella? crassa?** Sow. (Siluria, 2nd ed. pl. 9, figs. 6—8).	a. 240, D. G. Moel Seisiog.	
Ga 9		**Retzia Barrandii,** Davids. Mon. Sil. Brach. t. 13, figs. 10—13.	b. 577, W. L. Worcester Beacon.	
	Hemythyris crispata, p. 200.	**Rhynchonella Stricklandii,** var. **crispata** (Dav. Sil. Br. t. 21, fig. 28).	Old Radnor.	
Ga 7	*Hemithyris*, p. 201.	**Rhynchonella diodonta,** Dalm. Davids. Sil. Br. pl. 21, figs. 21—23.		b. 583, D. F. Balmae shore.
Ga	(On tablet with *Mytilus? unguiculatus.*)	**Rhynchonella 10-plicata,** Sow. (Siluria, 2nd ed. pl. 9, fig. 15).	a. 456, D. G. Plas Madoc, N. of Llanrwst.	
Ga 7	*Hemithyris*, p. 204.	**Rhynchonella nucula,** Sow. (Siluria, 3rd ed. pl. 9, fig. 9).		b. 585, D. F. Balmae shore.
Ga 8	*Hemithyris*, p. 203.	**Rhynchonella (Lewisii ?),** Davids. (Siluria, 3rd ed. foss. 58, figs. 1, 2).		b. 587, C. G. Middleton Fell.
Ga	*Hemithyris crispata*, p. 200.	**Rhynchonella,** a large species like *Stricklandii.*	a. 238, Old Radnor.	

Case and Column of Drawers.	Reference to McCoy's Synopsis: and Figures of Genera.	Names and References; Observations, &c.	Numbers and Localities.	
			Wales and Shropshire.	*Westmorland and S. of Scotland.*
G a 7		**Rhynchonella,** large species.		b. 589, C. F. Rebecca Hill, Ulverston.
		Pentamerus rotundus, Sow. (Dav. Sil. Brach. t. 15, figs. 9—12).		
G a 7	*P. microcamerus,* p. 210.	**Pentamerus liratus,** Sow. (*Stricklandinia,* Davidson, Mon. Sil. Brach. 1864, pl. 20).		b. 582, C. F. Rebecca Hill, Ulverston (Martin).
G a	p. 216.	**Orthis elegantula,** Dalm. Siluria, 2nd ed. pl. 20, fig. 12.	b. 592, D. G. Moel Seisiog; b. 593, Plas Madoc.	
G a		**Orthis,** sp. flat.	b. 594, D. G. Plas Madoc.	
G a 7		**Orthis,** new sp., like *biloba,* but larger.		a. 378, D. F. Balmae shore.
G a 6		**Orthis,** sp.? like *O. protensa,* Sow.		b. 595, C. F. Rebecca Hill, near Ulverston.
G a 6	*Lept. alternata,* p. 233.	**Strophomena,** sp.		a. 462, D. F. Balmae shore, Kircudbright.
G a 6	*S. compressa,* p. 242.	**Strophomena,** sp. fine lines.	b. 596, D. G. Moel Seisiog.	
G a 9	[As *S. euglypha,* p. 243.]	**Strophomena ouralensis,** De Vern. (*Leptæna,* Geol. Russ. Vol. II. p. 220, t. 14, fig. 1).	b. 597, W. L. Littlehope; b. 598, Old Radnor road to Presteign.	
G a 9	*Leptagonia depressa,* p. 248.	**Strophomena depressa,** Sow. (Siluria, 3rd ed. pl. 20, fig. 20).	b. 602, W. L. Littlehope, Woolhope.	
G a 6	*L. tenuissimestriata,* p. 239.	**Leptæna transversalis,** Dalm. (See Wenlock).	b. 599, D. G. Moel Seisiog.	
G a **G a 7**	*Leptæna,* p. 235.	**Chonetes lævigata,** Sow. (Siluria, 2nd ed. pl. 20, fig. 15).	b. 600, Gwyddelwern; b. 603, D. G. Devil's Bridge*.	
G a	do.	**Chonetes minima,** Sow. (Siluria, pl. 20, fig. 16).	b. 601, D. G. Gwyddelwern, Derwen.	

* Probably Devil's Bridge is but a doubtful locality for Denbigh Grits.

Case and Column of Drawers.	Reference to McCoy's Synopsis: and Figures of Genera.	Names and References; Observations, &c.	Numbers and Localities.	
			Wales and Shropshire.	*Westmorland and S. of Scotland.*
		LAMELLIBRANCHIATA.		
G a	p. 265.	**Mytilus ? unguiculatus** (Salter, (Mem. Geol. Surv. II. p. 365, pl. 20, fig. 6). A doubtful *Mytilus*, but near that genus.	a. 456, D. G. Plas Madoc.	
G a 6	Pl. 1 K, fig. 6, p. 286.	**Nuculites post-striatus,** Emmons, Pal. New York, t. 34, fig. 10.	b. 607, D. F. Gwydd-elwern, Derwen (as Caradoc).	
G a 6	*Sanguinolites,* Pl. 1 I, fig. 24, p. 277.	**Orthonotus decipiens,** McCoy, sp.		a. 916, D.F. Balmae.
G a G a 6	Pl. 1 K, figs. 2, 3, p. 283.	**(Arca) Palæarca ? Edmondiiformis,** McCoy. A small obtusely subquadrate shell (not *Arca*).	a. 455, D. G. Moel Seisiog; b. 608, D. F. near Llangynyw.	
G a	*Nucula levata,* Pl. 1 K, figs. 4, 5, p. 285.	**Ctenodonta anglica,** D'Orb. (Siluria, 2nd ed. pl. 23, fig. 10).	Near Llangynyw; a. 457, D. G. Plas Madoc.	
G a 6		**Ctenodonta,** sp.	b. 609, Llangynyw Rectory.	C. F. Helms Knot, Dent.
G a G a 6	p. 282.	**Cardiola interrupta,** Broderip (Sil. 2nd ed. pl. 23, fig. 12).	b. 611, D. F. Cefn ddu.	b. 612, C. F. Helms Side; b. 613, Troutbeck, Windermere; b. 614, C. G. Middleton Fell, near Kirkby Lonsdale, N. end.
G a 8	*Sanguinolites inornatus,* p. 277.	**Orthonota,** sp.		b. 615, D. F. Balmae shore.
G a	Pl. 1 K, figs. 7, 8, p. 273.	**Clidophorus ovalis,** McCoy.	a. 241, D. G. Plas Madoc.	
		Pterinea, Goldfuss, comprehends most of the *Aviculæ* of Silurian rocks.		
G a 6	Pl. 1 I, fig. 3, p. 263.	**Pterinea subfalcata,** McCoy.		b. 616, C. G. Howgill Fell, near Sedbergh.
G a G a 8	Pl. 1 I, fig. 4, p. 263.	**Pterinea tenuistriata,** McCoy. A very common and characteristic species of the Coniston grits.		b. 617, C. G. Above Ravenstonedale; b. 618, Middleton Fell, Kirkby Lonsdale; b. 619, C.F. Helms Side.

Case and Column of Drawers.	Reference to McCoy's Synopsis: and Figures of Genera.	Names and References; Observations, &c.	Numbers and Localities.	
			Wales and Shropshire.	*Westmorland and S. of Scotland.*
		GASTEROPODA.		
Ga 5		**Pleurotomaria** (or **Trochonema ?**) Two species large, turbinate.	a. 610, W. L. Little-hope.	a. 610, D. F. Balmae shore.
Ga 10		**Hormotoma (Murchisonia),** sp. Very slender whorls.		
Ga	p. 303.	**Holopella gregaria,** Sow. (Siluria, 2nd ed. pl. 34, fig. 10 a).	b. 620, D. G. Plas Madoc.	
Ga 9	Pl. 1 K, fig. 40, p. 296.	**Trochus? cælatulus,** McCoy, probably a *Cyclonema.* There are no species of true *Trochus* in Palæozoic rocks.	a. 460, W. L. Old Radnor road.	
		HETEROPODA.		
	p. 311.	**Bellerophon trilobatus,** Sow. (Siluria, 2nd ed. pl. 34, fig. 9). The common Bellerophon of sandy deposits in Silurian rocks: but chiefly common in Upper Ludlow rock.	D. G. Plas Madoc, and other localities near Llanrwst.	
Ga 10	Pl. 1 L, fig. 25, p. 311.	**Bellerophon subdecussatus,** McCoy. A pretty subglobate species, reminding us of Carboniferous forms.	a. 612, D. F. Llanrwst, Denbighshire.	
		CEPHALOPODA.		
Ga Ga 10 Ga 8	p. 317.	**Orthoceras subundulatum,** Portl. (Siluria, 3rd ed. Foss. 62, fig. 3).	b. 540, D. G. Moel Seisiog; D. F. Craig ddu allt.	b. 541, C. G. Helms Knot; b. 541*, Howgill Fell near Sedbergh. b. 542, C. F. Road from Coniston to Hawkshead; b. 542*, Cold Well, near the Castle; b. 543, Horton Dry Ridge.
Ga	p. 316.	**Orthoceras primævum,** Forbes sp. (Siluria, 3rd ed. Foss. 62, fig. 4).	b. 544, D. F. Cefn ddu.	b. 544*, C. F. Horton Dry Ridge.

Case and Column of Drawers.	Reference to McCoy's Synopsis: and Figures of Genera.	Names and References; Observations, &c.	Numbers and Localities.	
			Wales and Shropshire.	*Westmorland and S. of Scotland.*
G a	p. 318.	**Orthoceras ventricosum,** Sharpe (Quart. Journ. Geol. Soc. 2, pl. 13, f. 3).	b. 543*, D. F. Pen-y-Big, Bron Einion.	
G a 10		[These last two slight thin shells were regarded by Prof. Forbes as species of *Pteropoda* (*Creseis*), the septa are very seldom visible, being easily crushed.]	b. 545, D. F. Corwen Saw Mills; b. 545*, Llangynyw Rectory.	
G a 10	p. 313.	**Orthoceras angulatum,** Wahl. (Siluria, 2nd ed. pl. 28, fig. 4).	b. 540*, D. F. Llangynyw Rectory.	
G a 8		**Orthoceras,** sp.		b. 537, C. G. N. end of Middleton Fell, Kirkby Lonsdale.
G a 10 G a 8	p. 320.	**Orthoceras subannulatum,** Münst. Beiträge zur Petref. p. 99, t. 19, fig. 3.		b. 546, C. F. Road from Coniston to Hawkshead. b. 546*, C. G. Helmside.
G a 10		**Orthoceras,** sp.		b. 537*, C. F. Rebecca Hill, Ulverston.
G a G a 10	p. 317.	**Orthoceras tenuicinctum,** Portlock (Siluria, 3rd ed. Foss. 42, fig. 3).		b. 536, C. F. Road from Coniston to Hawkshead; b. 536*, Helmside.
G a 10	p. 315.	**Orthoceras laqueatum,** Hall? (Pal. New York, t. 56, fig. 1, p. 206).		b. 547, C. F. Road from Coniston to Hawkshead.
G a 10	p. 314.	**Orthoceras dimidiatum,** Sow. (Siluria, 2nd ed. pl. 28, fig. 5).		b. 547*, C. F. Helms Knot, Dent.
G a		**Orthoceras** (sp. smooth).		b. 532, C. F. Dry Ridge, Horton.
G a 8	p. 321.	**Orthoceras tracheale,** Sow. (Siluria, 2nd ed. pl. 34, fig. 6).		b. 532, C. G. Howgill Fell near Sedbergh.
G a 10	p. 319.	**Orthoceras annulatum,** Sow. (Siluria, 2nd ed. pl. 26, figs. 1, 2).	b. 539*, W. L. Old Radnor to Presteign.	
G a		**Trochoceras,** sp. Very ill preserved and filled with spar.		b. 559, C. F. Dry Ridge, Horton in Ribblesdale.

Wenlock Limestone and Shale.

It is not advisable to keep these separate, for in the Eastern Counties the limestone often dies away to a mere trace, and in S. Wales it is rarely to be seen at all. In North Wales and Westmorland the Wenlock formation is chiefly mud, slate, or shale. In S. Wales shale and sandstone, more frequently a clayey sandstone. The fossils consequently vary much in different localities, the more sandy strata often containing species which in Shropshire are confined to sandy beds in the Ludlow rock.

F. C. Fletcher Collection, part of which is in a separate cabinet F. C.

Case and Column of Drawers.	Reference to McCoy's Synopsis: and Figures of Genera.	Names and References; Observations, &c.	Numbers and Localities.
		Plantæ veræ. Most of the species known as *Fucoids* (i. e. seaweeds, fossil) are nothing more than the filled-up burrows of worms, &c.	
		Chondrites, Brongn. A genus intended to include many different forms.	
F C		**Chondrites verisimilis,** Salter (Expl. Edinb. Memoir, Geol. Surv. p. 134, pl. 11, fig. 1). A true sea-weed, like *Fucus cartilagineus.*	a 317, Dudley, F. C.
		AMORPHOZOA (*Sponges*).	
	Sponges.	**Stromatopora, Coscinopora, Cnemidium, Verticillopora, Stellispongia,** &c. are examples of very solid calcareous sponges. *Ischadites, Sphærospongia, Amphispongia,* and other Silurian forms are supposed to be distantly allied to the living *Grantia.*	
G b 10 F C	p. 12.	**Stromatopora striatella,** D'Orb. (Siluria, 2nd ed. Foss. 51, pl. 41, fig. 31). *S. Concentrica,* Lonsdale, Sil. Sys. t. 15, fig. 31 (not of Goldfuss).	a. 683, a. 684, Wenlock; b. 655, Dudley, F. C.; a. 684, good polished section.
F C		**Cnemidium tenue,** Lonsd. (Siluria, 2nd ed. pl. 38, fig. 11). An obscure genus of calcarcous sponges with minute *oscula.* Such sponges, and such as the following species:—	a. 318, Dudley, F. C.
G		**Verticillopora abnormis,** Lonsd. (Siluria, 2nd ed. Pl. 38, fig. 10), are common in the Wenlock Limestone here and there.	Dudley, F. C.
		(**Fistulipora,** McCoy. See *Corals.* I doubt if this be more than a sponge. J. W. S.)	

Case and Column of Drawers.	Reference to McCoy's Synopsis: and Figures of Genera.	Names and References; Observations, &c.	Numbers and Localities.
F C		**Ischadites Kœnigii,** Murch. (Siluria, 2nd ed. pl. 12, fig. 4). There is no doubt this is one of the regular sponges like *Sphærospongia* (p. 40). It has a root, and foramen at top, after the manner of *Grantia* (Bow.).	a. 33, Dudley (shale) F. C.; a 32, Malvern (shale) F. C.
F C G		**Sponge,** new. Palmate lobed.	a. 319, Dudley, F. C.
F		**Sponge,** cylindrical, vermiform, acrogenous in growth. This is quite new in Silurian rocks, and should be figured.	a. 320, Dudley, F. C.
		Some of these are doubtful, and are not catalogued by Prof. McCoy. Chiefly from Count Münster's collection, as *Cellepora,* &c.	
	Bryozoa.	**BRYOZOA or POLYZOA.**	
		Low polypoid Mollusca (lace corals, *Flustræ,* &c.) abound on every quiet sea bottom.	
F C		New form like *Archimedea* of the M. Limestone of America.	a. 321, Dudley, F. C.
G b 2		**Bryozoa,** sp. 1. Cells oblong, granular between.	a. 310, Dudley, F. C.
G b 2		**Bryozoa,** sp. 2. Thick branches, cells permanent.	a. 309, Dudley, F. C.
G b 2		**Bryozoa,** sp. 3. Encrusting on Favosites, granules very prominent.	a. 322, Dudley, F. C.
G b 2		**Trematopora** (Hall), sp. Branched cells in linear row. See Pal. N. York, Vol. II. p. 149.	a. 323, Dudley, F. C.
G b 2		**Discopora?** Rising over *Syringopora.* (Siluria, 2nd ed. pl. 41).	a. 885, Dudley, F. C.
G b 2		**Ceriopora?** Branched species (like Hornera in aspect). This may possibly be a coral allied to *Cœnites.*	a. 884, Dudley, F. C.
F C		**Fenestella infundibulum,** (*Retepora,* Lonsdale, Siluria, 2nd ed. pl. 41, fig. 24).	b. 656, Dudley, F. C.

Case and Column of Drawers.	Reference to McCoy's Synopsis: and Figures of Genera.	Names and References; Observations, &c.	Numbers and Localities.
F C G b	p. 49.	**Fenestella Milleri,** Lonsdale (Siluria, 3rd ed. Foss. 50, fig. 4, pl. 41, fig. 17). The old state is very rare indeed.	a. 327*, Dudley, F. C. old state.
F C	1 c, fig. 19, p. 50.	**Fenestella rigidula,** McCoy. *F. elegans,* Hall. Pal. N. York, Vol. II. pl. 40 D, fig. 1. The branches are ridged sharply: very parallel pores three between each bar.	a. 326*, Dudley, F. C. in large quantity.
F C	1 c, fig. 20, p. 50.	**Fenestella patula,** McCoy. More open meshes and more sinuous branches than last.	a. 325*, Dudley, F. C.
	p. 50.	**Fenestella subantiqua,** D'Orb. (Siluria, 3rd ed. pl. 41, fig. 16; Foss. 30, fig. 1).	Dudley, F. C.
F C		**Fenestella,** sp. 1. Coarse and very irregular branches.	a. 325, Dudley, F. C.
G b 2		**Fenestella,** sp. 2. Somewhat more funnel-shaped, and with prominent pores, two between each interstice.	a. 326, Dudley, F.C.
F C		**Fenestella,** sp. 3. Allied to *F. infundibulum.*	Dudley, F. C.
F C		**Fenestella assimilis,** Lonsd. Siluria, 2nd ed. pl. 41, fig. 27; Foss. 49, fig. 2.	b. 657, Dudley, F. C.
F C		**Dictyonema retiformis,** Hall (Pal. N. York, 2, t. 40. A horny form of the Retepora Group (not the same as Cambrian Fossil. J. W. S.).	a. 327, Dudley, F. C.
F C G b 2	p. 49. 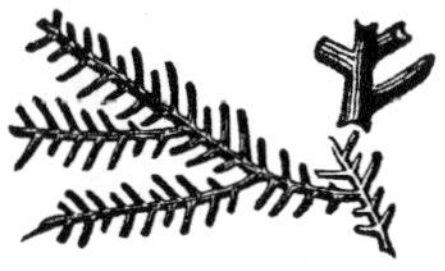	**Glauconome disticha,** Goldf. (Siluria, 2nd ed. pl. 41, fig. 12; Foss. 49, fig. 5). This seems to be everywhere characteristic of the Wenlock Limestone. A central stem and short lateral branches to each frond, and the fronds grow in clusters from one root like a herring-bone.	a. 393*, Dudley.
F C		**Glauconome,** sp. ? With obliquely set branches, otherwise the same as the last.	a. 393, Dudley, F. C.

Case and Column of Drawers.	Reference to McCoy's Synopsis: and Figures of Genera.	Names and References; Observations, &c.	Numbers and Localities.
		Ptilodictya, Goldf. Petref. Germanica. Flat plates, with the cells very regularly arranged on both sides of them. The species are often branched, once or twice, but always in one plane. The genus is world-wide in Cambrian and Silurian, and very numerous in species.	
F C G b 9	p. 47.	**Ptilodictya lanceolata,** Goldf. (Pet. Ger. t. 37, fig. 2) Siluria, 2nd ed. Foss. 49, fig. 6, pl. 41, fig. 11. Often attains great size, ten inches; easily breaks along the centre plate.	Aymestry (Haven), b. 658, Dudley, F. C.; b. 661, Ledbury.
F C		**Ptilodictya scalpellum,** Lonsd. (Siluria, 2nd ed. Foss. 50, pl. 41, fig. 25). Slightly branched only: but leading away from the compound branched and explanate species of the Cambrian to the simple *P. lanceolata.* Such species as the present would be included in Prof. Hall's unnecessary genus *Stictopora.*	b. 659, Dudley, F. C.
F C G b 9		**Polypora or Hornera crassa,** Lonsd. (Siluria, 2nd ed. Foss. 49, fig. 1, pl. 41, fig. 13). *Hornera* only differs from *Polypora* (Devonian and Carboniferous) by wanting the connecting bars.	b. 660, Dudley, F. C.
		HYDROZOA? or POLYZOA.	
		The Graptolites are (I think not rightly) generally now referred to the *Hydroid Polypes.* I prefer the analogy with such Bryozoa as *Defrancia.*	
G b 1	*G. ludensis,* p. 4.	**Graptolites priodon,** Bronn (*G. ludensis*) (Siluria, 2nd ed. pl. 12, fig. 1, Foss. 11, fig. 3). See remarks on this common Northern fossil in the Cambrian lists. It is extremely common at the base of the Wenlock in France and Bohemia.	a. 468, S. of the Dee, Llangollen; b. 662, Smithfield; b. 663, Cwmbach, Builth; also b. 664, bed of Wye, Builth; b. 665, Cwm, W. of Cefn Grugos, Llanfyllin; b. 666, Llanfair; b. 667, Ffyrnwy, Montgomeryshire.

Case and Column of Drawers.	Reference to McCoy's Synopsis: and Figures of Genera.	Names and References; Observations, &c.	Numbers and Localities.
G b 2		**Graptolites,** sp. With simple not recurved cell-mouths. Very common. It is possibly *G. colonus,* Barrande, in Lower Ludlow, but more likely distinct.	Wenlock shale? b. 668, Dudley, F. C. It is possible this comes from the Lower Ludlow shale, above the workable limestones. The species abounds near Ludlow.
G b 10	*Star corals, Millepores, &c.* (*True Corals*) with stony bases to cells.	**ZOOPHYTA (CŒLENTERATA,** Huxley). *Tabulata,* or Millepore Corals.	
	p. 11.	**Fistulipora decipiens,** McCoy (Ann. Mag. N. Hist. 2nd Ser. Vol. VI. p. 285). The cell tubes are so minute, and the interstitial substance too like a sponge.	
	Palæopora, McCoy. p. 14.	**Heliolites,** Dana. A common Wenlock fossil. Millepore corals allied closely to the *Heliopora cœrulea* of V. D. Land. Cell tubes set in a mass of *cœnenchyma,* i. e. pointed small tubes of much less diameter than the cells. Species numerous.	
G b 3 F C G b 3	*Palæopora,* p. 15.	**Heliolites interstincta,** Wahl. (Siluria, 2nd ed. Foss. 18, figs. 3—5, pl. 39, fig. 2. Milne Edw. t. 57, fig. 9). Incrusting on shells (*Orthoceras,* or spiral shells chiefly) or amorphous or pyriform when grown in rapidly accumulating mud. In fact, it assumes all shapes, like a *Proteus,* according to circumstances, and ranges from Middle Cambrian to Ludlow rock.	b. 671, Wenlock Ridge, Woolhope; b. 673, Malvern; b. 674, Aymestry; b. 675, Dudley, F. C.; b. 675*, Dudley, var. F. C. b. 672, Walsall, var. F. C.
F C	*P. subtubulata,* p. 16?	**Heliolites,** var. **Murchisoni,** Milne Edw. Brit. Foss. Cor. tab. 57, fig. 6.	b. 676, Dudley, F. C.
F C G b 3	Pl. 1 c, fig. 4. *Palæopora,* p. 16.	**Heliolites megastoma,** McCoy (Sil. Foss. Irel. pl. 4, fig. 14). Siluria, 2nd ed. 1859, woodcut 27, fig. 7. Easily known from the last (not so common) by the very large close set cups. (Milne Edw. t. 58, fig. 2).	b. 677, Wenlock Ridge, Woolhope; b. 678, Dudley, F. C.
F C	*Palæopora,* p. 17.	**Heliolites petalliformis,** Lonsd. (*Plasmopora,* Milne Edw. Brit. Foss. Cor. p. 253, pl. 59, fig. 1. Siluria, 2nd ed. Foss. 18, fig. 2, pl. 39, fig. 4). Regarded as a distinct genus (*Plasmopora*) by Milne Edw. The distinctions seem too minute to be more than specific.	b. 680, Dudley, F. C.

Case and Column of Drawers.	Reference to McCoy's Synopsis: and Figures of Genera.	Names and References; Observations, &c.	Numbers and Localities.
F C		**Heliolites,** var. With depressed areas at the cell mouths.	b. 681, Dudley, F. C.
F C G b 3	*Palæopora,* p. 18.	**Heliolites tubulata,** Lonsd. (Siluria, 2nd ed. pl. 39, fig. 3), and Foss. 18, fig. 1.	b. 682, Dudley, F. C.; b. 683, Wenlock; b. 684, Aymestry; b. 685, Woolhope.
F C		**Heliolites Grayi,** Milne Edw. (Brit. Foss. Cor. p. 252, t. 58, fig. 1, Archiv. Mus. v. p. 217). Flat branches; crateriform cells, raised edges.	a. 342, Dudley, F. C.
		Heliolites, var.? With sunk pores.	
F C		**Heliolites cæspitosa,** n. s. Salter. Something like *H. Grayi,* but cells approximate; shallow pits, not raised.	a. 343, Dudley, F. C.
F C		**Heliolites,** sp.	a. 376, Dudley, F. C.
	p. 19.	**Favosites** (*Calamopora,* Goldfuss). A genus of Silurian and Devonian corals—more abundant here than any other. The cells are thickly pierced with little round holes (*foramina*) often in two rows, but sometimes more, by which a communication is kept up between neighbour corallites. Close-set tabulæ fill up the hexagonal tubes, and there are scarcely visible septa.	
G b 4		**Favosites,** sp. Shewing the upper crowded tabulæ.	a. 369, Dudley.
G b 2	p. 19.	**Favosites alveolaris,** Goldf. (Petref. G. t. 26, fig. 1; Siluria, 2nd ed. Foss. 17, fig. 4). [I do not know how externally to distinguish this species from *F. gothlandica,* and I believe the *F. alveolaris,* with large serrated angles to the tubes, to be very common in our Wenlock Rocks.]	b. 685, Wenlock Ridge, Woolhope; b. 686, Dudley.
F c G b 4 G b	p. 20.	**Favosites Gothlandica,** Linn. (Goldf. Pet. Germ. t. 26, fig. 3; Siluria, 2nd ed. Foss. 17, figs. 2, 3, pl. 40, figs. 3, 4, Foss. 29, fig. 6). One of the most beautiful of fossils. The honey-comb cells when perfect are radiated by six obtuse folds which represent septa. The sides of the tubes are pierced by a regular double row of holes, except in the next *var.*	b. 687, Ledbury; Woolhope; a. 675, a. 676, Wenlock, tubes well-shewn; b. 688, Dudley.

Case and Column of Drawers.	Reference to McCoy's Synopsis: and Figures of Genera.	Names and References; Observations, &c.	Numbers and Localities.
G b 4	p. 21.	**Favosites,** id. var. **basaltica,** Goldf. (ib. pl. 26, fig. 4), which has only a single row of foramina. In all the varieties (and I suspect the following species is only the young state) the angles of the tubes are smooth.	b. 689, Dudley, labelled by Count Münster as *Calamopora basaltica.*
F C G b		**Favosites Forbesii,** Milne Edw. (Brit. Foss. Cor. t. 60, fig. 2, p. 258). Distinguished from *Fav. Gothlandica,* of which probably it is the young.	a. 681, Dudley, F. C., a. 400, F. C., a. 682, F. C., shews tubes inside full of granules: these are not the *foramina,* though often mistaken for them.
G b		**Favosites favosa,** Goldf. sp.? (Petref. Germ. t. 26, fig. 2). Only one specimen. The walls very thick and cells very large.	a. 464, Whitfield, Tortworth (Earl Ducie).
F C G b 4	p. 20.	**Favosites aspera,** D'Orb. (Milne Edw. and Haime, Brit. Foss. Cor. t. 60, fig. 3, Siluria, 2nd ed. pl. 40, figs. 1, 2).	a. 680, Dudley, F. C., young growths on good upper surface; a. 679, good, Dudley, F. C.; Wenlock; Aymestry; Malvern; a. 368, Whitfield, Tortworth (Earl Ducie).
G b 4		**Favosites Hisingeri,** Milne Edw. (Brit. Foss. Cor. t. 61, fig. 1, p. 259). I am not sure of this species: nor is any one else. It is probably a state of *F. aspera,* and both are probably varieties of *F. multipora,* Lonsdale, a species supposed to have many rows of pores: but certainly it has not—the pores are few. [J.W.S.]	b. 690, Woolhope hills.
F C G b 4		**Favosites cristata,** Blumenbach (Milne Edw. and Haime, t. 61, fig. 3; Siluria, 2nd ed. Foss. 17, fig. 1, pl. 41, fig. 2). It is very variable in shape and size of cells, and is much like *F. polymorpha,* the Devonian species. But it has greatly thicker walls to the cells: one species was referred to *F. polymorpha* formerly.	a. 398, Dudley. a. 398*, Wenlock; a. 399, Dudley, F. C., both varieties, branches narrow and large, and small and large-celled forms.
	p. 21.	**Cœnites,** Eichwald (*Limaria,* Londs.). Branched, or lobed, like *Favosites,* but with thick walls not perforated—or not known to be so—and with oblique mouths, which are much thickened so as to leave but a slit or triangular opening.	

Case and Column of Drawers.	Reference to McCoy's Synopsis: and Figures of Genera.	Names and References; Observations, &c.	Numbers and Localities.
G b 5 F C	p. 22.	**Cœnites intertextus,** Eichwald (Milne Edw. t. 65, fig. 5. *Limaria fruticosa,* Sil. Syst. *Cœnites,* Siluria, 2nd ed. pl. 38, fig. 8).	b. 691, Sedgley? Aymestry; a. 300*, Dudley, F. C.
G b 5	Pl. 1 c, fig. 8, p. 22.	**Cœnites strigatus,** McCoy. Distinguished by Prof. McCoy by the scratch-like channels of the worn cell-mouths, but probably *C. intertextus.*	b. 692, Dudley.
G b 5 F C		**Cœnites juniperinus,** Eichw. (Milne Edw. t. 65, fig. 4. Siluria, 2nd ed. Foss. 19, fig. 3. *Limaria clathrata,* pl. 38, fig. 7).	b. 693, Dudley, F. C.; a. 299, Dudley, F. C.
F C		**Cœnites labrosus,** Milne Edw. t. 65, fig. 6, p. 277. A wide flat-lobed species with short mouths to the cells, with projecting lips or edges. The mouths rather open. Occasionally the projection is small.	Dudley, F. C.
F C		**Cœnites linearis,** Milne Edw. (t. 65, fig. 3, p. 277). Like the last, but with linear mouths on the projecting bosses.	Dudley, F. C.
F C		**Cœnites,** sp. 1. Like *linearis,* but with no depth of mouth. Cell-openings curved, shallow, but very low set.	a. 301, Dudley, F. C.
F C		**Cœnites juniperinus,** var. Closer mouths, and somewhat longer.	Dudley, F. C.
F C		**Cœnites,** sp. 2. Allied var. to *linearis,* but with crested-edged mouths. This character in extreme, as in sp. 1. The extreme of smoothness is gained.	a. 300, Dudley, F. C.
F C		**Cœnites,** sp. (a var. of *C. labrosus*). With no projections to support the mouth.	Dudley, F. C.
	p. 68.	**Alveolites,** Lonsd. The mouths of the cells are oblique or rhomboidal, but not thickened, and there is generally a thickened ridge or tooth on one side; tabulæ imperfect; septa none, unless the ridge above named be one. Range—Silurian to Carboniferous.	

Case and Column of Drawers.	Reference to McCoy's Synopsis: and Figures of Genera.	Names and References; Observations, &c.	Numbers and Localities.
		A. *Branched Species* (*Cladopora*, Hall).	
G b 2 F C G b 4	*Favosites oculata*, p. 21.	**Alveolites repens,** Linn. (Milne Edw. t. 62, fig. 1, p. 263. Siluria, 2nd ed. Foss. 17, fig. 6). *Favosites oculata,* Goldfuss (Petref. Germ. t. 65, fig. 14), according to McCoy.	b. 694, Wenlock; a. 397, Dudley, F.C.; a. 367, Dudley, F.C., a choice specimen shewing various states of the surface.
F C		**Alveolites Fletcheri** (Seeley MSS.), allied to *A. repens.* The branches frequently connate and lobed. Narrow branches: small cell-mouths.	a. 394, Dudley, F. C.
G b 4 F C		**Alveolites seriatoporoides,** Milne Edw. (Brit. Foss. Cor. t. 62, fig. 2, p. 263).	b. 695, Dormington; b. 696, Wenlock Edge; a. 398, Dudley, F. C.
		B. *Amorphous* or *round forms.*	
G b 4 F C		**Alveolites Labechii,** Lonsd. (Siluria, 2nd ed. pl. 40, fig. 8. Foss. 17, fig. 5). The sp. occurs in various forms, flattened, round, or irregularly expanded.	b. 697, Woolhope quarries; one specimen is figured by M. Edwards, a. 396, as *Monticulipora Bowerbankii.*
G b 7		**Alveolites Seeleyi,** Salter, n. sp. Branches coalesced, cells small. Distinguished by the acumen of Mr Seeley.	a. 365, Dudley, F. C.
		Stenopora, Lonsd. The genus is a little doubtful: it depends upon the contraction of the mouth of the cells, and on the irregular tabulæ. The species are not easily distinguished from *Chætetes* or *Monticulipora.*	
G b 5 F C G b 3	p. 24.	**Stenopora fibrosa,** Goldf. (Pet. Germ. t. 28, fig. 3). Siluria, 2nd ed. Foss. 17, fig. 7, Foss. 30, figs. 1, 2, pl. 40, figs. 6, 7. Easily distinguished from *Monticulipora* by the structure of tubes.	Falfield or Whitfield, Tortworth (Earl Ducie); b. 698, Tottlebank, near Ulverston; a. 377, Dudley, F. C.; a. 891, Dudley, Münster Coll.
	p. 26.	**Stenopora granulosa,** Goldf. (*Ceriopora*). Is a *Chætetes.* See that genus.	

Case and Column of Drawers.	Reference to McCoy's Synopsis : and Figures of Genera.	Names and References ; Observations, &c.	Numbers and Localities.
		Chætetes Fischer. A genus of *Tabulata* which is very like *Stenopora,* and *Monticulipora* especially, but has not the clustered fertile pores of the last. Tubes increase not by lateral gemmation, but by fission of tubes, which are inseparably united ; hence in splitting, the interiors (not the walls) shew. No foramina on the walls.	
F C		**Chætetes,** sp. With long, narrow, linear branches, very variable in diameter and length.	a. 401, Dudley, F. C.
		Chætetes, sp.	
F C G b 5		Several other species remain to be described. Some may be *Monticulipora* (most probably are so).	Dudley, F. C.
	p. 24.	**Nebulipora,** McCoy. **Monticulipora,** Milne Edwards. Amorphous or branched corals, or even encrusting, with clustered cells (fertile) of larger size than the ordinary barren ones—and these fertile cells frequently on prominences.	Dudley, F. C.
		1. Branched species.	
F C		**Monticulipora Fletcheri,** Siluria, 2nd ed. pl. 40, fig. 9. M. Edwards (t. 62, fig. 3). Walls thick, cells prominent, pores (interstitial) minute, evidently the young state of *Pulchellus.*	a. 366*, Dudley, F. C.
F C		**Monticulipora pulchella,** M. Edwards (t. 62, fig. 5), p. 267, rather coarse pores (prominent edges), but not angular pores.	a. 402, Dudley, F. C. (fig. specimen).
F C G b 5		**Monticulipora,** sp. 1. Thick branches, rather small. Angular cells, very distinct clusters.	a. 305, Dudley, F. C., common.

Case and Column of Drawers.	Reference to McCoy's Synopsis: and Figures of Genera.	Names and References; Observations, &c.	Numbers and Localities.
F C		**Monticulipora Bowerbankii,** Milne Edwards (t. 63, fig. 1), p. 268. Coarse angular pores, and not very distinct or separate clusters, often much branched. *Nebulipora Bowerbankii,* Siluria, Appendix, 534.	a. 306, Dudley, F. C. is the figured (1 b.) species of Edwards.
G b 6 G b 5		**Monticulipora,** sp. Smaller cells, and of more globular form than *M. Bowerbankii.*	Wenlock Edge, Woolhope. a. 308, Whitfield (Earl Ducie).
		2. *Papillate* species. Incrusting or expanded.	
F C	Pl. 1 C, fig. 5, p. 24.	**Monticulipora papillata,** McCoy (M. Edw. t. 62, fig. 4). True sp. of McCoy. Small angular cells, sharp prominent bosses. (Incrusting.)	a. 311, Dudley, F. C.
F C		**Monticulipora,** sp. 2. Like *papillata,* but larger cells, and solid, scarcely raised papillæ. (Incrusting.)	a. 312, Dudley, F. C.
F C		**Monticulipora,** sp. 3. Cells minute, but prominent, (thick walls), papillæ solid. [Expanded 2 or 3 inches.]	a. 313, Dudley, F. C.
		3. *Lenticular masses.*	
F C		**Monticulipora poculum,** Salter, MSS.	a. 314, Dudley, F. C.
F C		**Monticulipora,** sp. 4. Cup-shaped base, upper surface therefore hollow, and sometimes nearly flat, with angular cells. But this is an unusual form in the genus. Sometimes the species is flatter.	a. 315, Dudley, F. C.
F C		**Monticulipora,** sp. 5. Gibbous, cells large, angular; the fertile cells scarcely larger than the rest, in small few-celled clusters.	a. 316, Dudley, F. C.
F C		**Monticulipora,** sp. 6. Like last, but much thicker walls and smaller cells, which are round not angular.	a. 317, Dudley, F. C.
	For figures of the genera of Corals see Milne Edwards' excellent work, *British Fossil Corals* (Palæontographical Society).	**Labechia,** Milne Edwards and Haime. Flat expanded corals, very unlike the usual tabulated forms, except *Thecia.* The surface is covered with elevated papillæ.	

Case and Column of Drawers.	Reference to McCoy's Synopsis: and Figures of Genera.	Names and References; Observations, &c.	Numbers and Localities.
F C G b 6		**Labechia conferta,** Lonsdale (Milne Edwards, Brit. Foss. Cor. p. 269), Siluria, 2nd ed. pl. 39, fig. 5.	a. 297, Dudley, F.C.; b. 701, Wenlock Edge, Woolhope.
G b 2		**Labechia** sp. Or var. of the above.	a. 298, Dudley, F. C.
		Thecia, Milne Edwards. An expanded coral like *Labechia,* with cell-mouths, almost closed by the thick sector, which all but fills up the tube. In this respect it resembles the modern star-corals, and perhaps indicates a passage to them. The section is very solid, only two species are certainly known. One is common in our Wenlock rocks.	
F C G b		**Thecia Swindernana,** Goldfuss. (Pet. Germ. pl. 38, fig. 3, Milne Edwards, t. 65, fig. 7). Variable in the size of cells.	b. 704, Dudley, F.C.; Woolhope; b. 703, Wenlock.
	Palæopora, p. 14.	**Thecia (Porites) expatiata,** Lonsd. (Siluria, 2nd ed. pl. 41, fig. 3).	
F C		**Thecia Grayana,** (Milne Edwards, t. 65, fig. 8), with fewer septa.	a. 403, Dudley, F. C.
F C		**Thecia,** species probably new. This form—separated by Mr Seeley in 1865, as probably distinct—seems to differ chiefly in the crowded? calices, and the consequent depth of the cups.	
F C		**Thecia Grayana,** var.? Lobed, almost palmate.	a. 302, Dudley, F. C.
G b 2		**Tabulate coral,** with linear connecting ridges, ? genus.	a. 303, Dudley, F. C.
	p. 26.	**Halysites** (chain coral). It differs from all other tabulate corals by having the calices connected fully by their sides, on two edges only, so as to form linear rows, which inosculate and form a reticulate pattern. The calices are oval, a rare circumstance, except in *Alveolites* before described. A section shews the tabulæ very close set; and there are no septa whatever or trace of them.	

Case and Column of Drawers.	Reference to McCoy's Synopsis: and Figures of Genera.	Names and References; Observations, &c.	Numbers and Localities.
F C G b 6	p. 26.	**Halysites catenularius,** Linn. (*Catenipora escharoides,* Siluria, 2nd ed. pl. 40, fig. 14, Goldfuss, Pet. Germ. t. 25, fig. 4). The commonest of common fossils, in every region where Upper Silurian is known. Arctic circle, &c.	a. 370, Dudley, F. C.; b. 705, Woolhope.
F C	do.	**Halysites,** var. **labyrinthicus,** Goldfuss, is only a variety with very large tubes.	a. 351, Dudley, F. C.
	p. 27.	**Syringopora,** Goldfuss. This is perhaps the extreme form which could be taken by Millepores, the tubes or corallites being absolutely separate except at the creeping base. They throw out lateral appendicles which unite and help to support the reticular mass. The base is known as *Aulopora,* Goldfuss.	
F C		**Syringopora fascicularis,** Linn. (*S. filiformis,* Lonsd.? Goldfuss, Pet. t. 38, fig. 6, Milne Edwards, t. 65, fig. 1), the most slender tubed of British Silurian species. *Aulopora tubæformis,* Lonsd. Sil. Syst. (Siluria, 2nd ed. pl. 40, fig. 12, pl. 41, fig. 8).	a. 352, Dudley, F. C.
F C G b 6 G c	p. 27.	**Syringopora bifurcata,** Lonsdale, *S. reticulata,* Sil. Syst. t. 15 *bis,* f. 11 (Siluria, 2nd ed. pl. 40, figs. 10, 11). A very large-tubed species, the largest indeed, and with branches most remote. Siluria, 2nd ed. Foss. 19, figs. 2, 4, 5.	b. 708, Wenlock Edge, Woolhope; b. 707, Dudley; a. 353, Dudley, F. C.
F C		**Syringopora serpens,** Linn. (M. Edw. t. 65, fig. 2). *Aulopora serpens,* and *conglomerata.* Lonsd. Siluria, 2nd ed. pl. 41, figs. 6—9.	a. 354, Dudley, F. C.
	Cup-corals.	**ZOANTHARIA RUGOSA,** M. Edwards and Haime. The cup and star-corals of the Silurian rocks are found to have more of the character of the *Tabulata* than any recent reef-building or cup-corals; and as a very remarkable character, the number of the septa is either four or a multiple of four, a character only found in the *Alcyonarian* Zoophytes of the present seas (Edwards and Haime).	

Case and Column of Drawers.	Reference to McCoy's Synopsis: and Figures of Genera.	Names and References; Observations, &c.	Numbers and Localities.
F C		**Aulacophyllum mitratum,** Hisinger (Leth. Suec. p. 100, Milne Edwards, t. 66, fig. 1). A compressed and very oblique cup. There is some doubt if the large form be not quite distinct from the small one.	a. 328, a. 329, Dudley, F. C.; a. 329 is fig. 1, a. 328 is fig. 1a in M. Edwards' Monograph; fig. 1 a. is the large form.
	p. 69.	**Cyathophyllum,** Goldfuss. The common Palæozoic genus of cup-corals.	
G b 7		**Cyathophyllum Loveni,** Milne Edwards (Brit. Foss. Cor. t. 66, fig. 2, p. 280).	a. 364, Dudley, F. C.
F C		**Cyathophyllum.** Large variety, possibly new species.	Dudley, F. C.
F C G b 8	*Strephodes,* Pl. 1 B, fig. 20, p. 30.	**Cyathophyllum pseudoceratites,** McCoy sp. M. Edw. (Brit. Foss. Cor. t. 66, fig. 3). **Cyathophyllum angustum,** Lonsd. See *C. articulatum,* Siluria, 2nd ed. pl. 39, fig. 9.	a. 333, Dudley, F.C., figured specimen of M. Edwards; a. 372, Sedgley, McCoy's figure.
G b 7		**Cyathophyllum.** Tufted narrow species like *C. flexuosum.*	a. 362, Dudley, F. C.
G b 8 F C	*Strephodes vermiculoides,* p. 31.	**Cyathophyllum truncatum,** Linn. (Milne Edw. t. 66, fig. 5, Siluria, 2nd ed. pl. 39, fig. 12, *b, c, d, Cyathophyll. dianthus,* Siluria, pl. 39, fig. 12). Easily known by the open flat limb of the cup and the calycular gemmation, 7 or 8 young buds often spring from one cup.	Lindell's Quarry, Woolhope. a. 330, Dudley, F.C.; a. 373, Wenlock.
G b 8 F C	*Strephodes vermiculoides.* Pl. 1 B, fig. 22, p. 31.	**Cyathophyllum articulatum,** Wahl. (M. Edw. Brit. Foss. Cor. t. 67, fig. 1, Siluria, 2nd ed. pl. 39, fig. 10, *C. cæspitosum,* Sil. Syst. pl. 16, fig. 10). Like the last, but much more slender stems.	b. 706, near Aymestry, McCoy's figure; a. 331, Dudley, F. C. (large figured specimen, Milne Edwards).
F C		**Cyathophyllum flexuosum,** Lonsd. (Milne Edw. t. 67, fig. 2. Siluria, 2nd ed. pl. 39, fig. 7). Usually a straight slender tube: it is an exception to find it curved as in Lonsdale's figure.	a. 332, Dudley, F. C.

Case and Column of Drawers.	Reference to McCoy's Synopsis: and Figures of Genera.	Names and References; Observations, &c.	Numbers and Localities.
F C G c	*Strephodes,* Pl. 1 B, fig. 21, p. 31.	**Cyathophyllum trochiforme,** McCoy, distinguished by the proportions of the lamellæ and the cup.	a. 360, Dudley, figured specimen, McCoy; a. 334, Dudley, F. C.
F C G b 7		Several sp. undescribed of this genus, or rather group of genera, remain in the drawers for future work.	Dudley, F. C. and Walsall, F. C. Also Woolhope.
G b 8		**Petraia,** sp. Probably a *Cyathophyllum,* as indicated by Milne Edwards and Haime.	Falfield, Tortworth (Earl Ducie).
		Omphyma, M. Edwards and Haime. A group of large *Cyathophyllidæ,* the largest (except *Ptychophyllum*) known in the Lower Palæozoic rocks: the cup is broad, and the septa do not reach the centre, which is occupied by broad flat tabulæ. The sides filled with the ordinary vesicular tissue. The outer coat (exotheca) generally sends down rootlets for support. Probably there are only two species in Britain. It is common in Sweden.	
F C G b 6 G c	*Caninia lata,* p. 28. Pl. 1 C, fig. 12, and *C. turbinata.*	**Omphyma turbinata,** Linn. (Milne Edwards, t. 69, fig. 1, Siluria, 2nd ed. Foss. 53, figs. 4, 5, pl. 39, fig. 11). The widest form of the genus, but the specimens vary so much, that it is possible the following may be only a variety (*C. lata* McCoy is certainly so).	a. 674, Dudley, F. C.; b. 711, Ledbury; b. 712, Woolhope (Dormington) quarries.
F C		**Omphyma subturbinata,** Milne Edwards, t. 68, fig. 1, differs chiefly in its greater length and the consequent narrowness of the cup.	b. 709, Dudley, F. C.
F C		**Omphyma ? Murchisoni,** Milne Edwards, t. 67, fig. 3. Siluria, 2nd ed. pl. 38, fig. 2.	b. 710, Dudley, F. C.
		Ptychophyllum, Milne Edwards and Haime. A large expanded cup-coral: the limb of the cup is so much turned down, and the columella or central part so much thickened and twisted, that the coral, which has but a very short stem or cup, seems turned inside out. The genus is not confined to	

Case and Column of Drawers.	Reference to McCoy's Synopsis: and Figures of Genera.	Names and References; Observations, &c.	Numbers and Localities.
		Britain. An Arctic form from the northern expeditions is in the cabinet of the Geol. Survey. In this the successive growths of the calyx are so regular, and the limb turned so much back, that the whole resembles a Chinese pagoda, whence the name *Ptychophyllum pagoda*, Salter.	
		Ptychophyllum patellatum, Schloth. (*Strombodes plicatum,* Lonsd., Siluria, 2nd ed. Foss. 52, fig. 5, pl. 38, fig. 4).	In every Wenlock collection.
		Goniophyllum, Milne Edwards and Haime. Square or semi-cylindrical coral-cups, of small size, not above an inch or two in height; with obscure septa, arranged in crucial fashion. The great peculiarity of this fossil (probably the remnant of a large class of extinct forms) seems to be that all its species, of which several have been described by Lindström from Sweden, are furnished with an operculum (*a*) or lid—and it is difficult to see how the tentacles performed their duty. *Calceola,* the Devonian fossil, hitherto thought a Brachiopod, is of this group.	
F C G c		**Goniophyllum pyramidale,** Hisinger, (Leth. Suec. p. 101. *G. Fletcheri,* Milne Edwards, t. 68, fig. 3, p. 290; *G. Fletcheri,* Salter in Siluria, 2nd ed. p. 244). It varies much in shape; the operculum has not been found yet in England, but should be looked for.	a. 335, Dudley, F. C.; Malvern, Wenlock Shale; b. 713, Dudley (J. Gray, Esq.).
		Compound cup-corals.	
		Only differ from single cups by the aggregate growth of the buds, or young corallites, which do not fall off the parent, but are attached and grow hexagonal by compression amongst one another.	

Case and Column of Drawers.	Reference to McCoy's Synopsis: and Figures of Genera.	Names and References; Observations, &c.	Numbers and Localities.
F C **G c**	*Acervularia ananas,* p. 35.	**Arachnophyllum luxurians,** Edw. and Haime, (Brit. Fos. Cor. t. 69, fig. 2, p. 292. Siluria, 2nd ed. pl. 39, fig. 6. *A. ananas,* Linn. (in part). Lonsdale, Sil. Syst. pl. 16, fig. 6. Siluria, 2nd ed. Foss. 53, fig. 6). There is no doubt this is the chief coral intended by *Linnæus,* but as he included more than one, the *A. ananas* was described by Lamarck from the other species.	a. 336, Dudley, F. C. is fig. 2*a* in M. Edwards' Monograph; a. 678 shews the buds well.
	p. 37.	**Arachnophyllum,** McCoy (*Strombodes,* Milne Edwards). McCoy's genus was certainly described shortly before or immediately after Milne Edwards' first description of the genus; and it seems hardly worth while to revive the doubtful genus of Schweigger, and displace a most excellent and graphic name for it.	
F C **G c** **G b 7** **G b 8**	Pl. 1 B, fig. 27, p. 38.	**Arachnophyllum typus,** McCoy (Milne Edwards, t. 71, fig. 1, as *Strombodes,* Siluria, 2nd ed. Foss. 52, fig. 6). *S. typus, diffluens, Murchisoni, Phillipsi,* and *Labechii,* all are varieties of one common species.	b. 716, Dormington, Woolhope; a. 361, near Aymestry, McCoy's figure; a. 363, Dudley, F. C. as *Murchisoni.*
	Pl. 1 B, fig. 28, p. 34.	[**Strombodes Wenlockensis,** McCoy, is a *Lonsdaleia,* Edwards, and is a common mountain limestone coral introduced by mistake; it never came from the Wenlock localities. J. W. S. 1867.]	From Shropshire or N. Wales.
	Sarcinula, p. 36.	**Syringophyllum,** Edwards and Haime (*Sarcinula,* Linn. in part). The projecting edges of the cups are the inner wall (endotheca), the interstices being filled by the coronate septa, and there are no bounding walls to the separate corallites—they are fused together as in modern reef-corals.	
	p. 37.	**Syringophyllum organum,** Linn. (*Sarcinula,* Lonsd. and Goldfuss, Pet. Germ. t. 24, fig. 10). More common in the Lower Silurian rocks. (See Milne Edwards, t. 71, fig. 3, and Siluria, 2nd ed. Foss. 29, fig. 4.)	No specimen; but should be obtained from Dudley.

Case and Column of Drawers.	Reference to McCoy's Synopsis: and Figures of Genera.	Names and References; Observations, &c.	Numbers and Localities.
		Vesicular cup-corals.	
	p. 32.	**Cystiphyllum,** Lonsd. In this cup-coral the septa, though not absent, are quite subordinate to the vesicular structure, which takes the place also of the regular tabulæ, or rather represents them.	
G b 8		A fine section to shew structure.	
F C		**Cystiphyllum Grayii,** Milne Edwards, t. 72, fig. 3. The original figured specimens are in this collection and are marked.	a. 337, Dudley, F. C.
G b 6	Pl. 1 B, fig. 19, p. 32.	**Cystiphyllum brevilamellatum,** McCoy. A somewhat doubtful species. McCoy seems to think Lonsdale's pl. 16 *bis*, fig. 2, Sil. Syst. may be this species.	a. 371, Wenlock, figured specimen.
G b 6 G b 7 F C		**Cystiphyllum cylindricum,** Lonsd. (Milne Edw. Brit. Fos. Cor. t. 72, fig. 2, also Siluria, 2nd ed. pl. 38, fig. 3).	Dudley, F. C.; b. 719, Wren's nest; b. 720, Woolhope (Wenlock Edge).
F C		**Cystiphyllum Siluriense,** Lonsd. (Siluria, 2nd ed. pl. 38, fig. 1). This wide species frequently has accessory roots like the *Omphyma* (p. 113).	a. 359, Dudley, F. C.
F C		**Cystiphyllum,** new sp. Compressed, and largely vesicular, one specimen shews the growth of seven buds from the calyx.	a. 338, Dudley, F. C.
		Palæocyclus, Milne Edwards and Haime. A genus of pretty discoid corals, of very small size; which from their external character were long regarded by Edwards and Haime as representatives of the *Fungidæ* or mushroom corals, and as the only coral of the order *Aporosa* in Silurian times. It really belongs to the *Rugosa* (Duncan).	
F C		**Palæocyclus porpita,** Linn. (Milne Edwards, Brit. Fos. Cor. t. 57, fig. 1). A circular and very depressed money-like species. Very common in the Dudley limestone.	a. 339, Dudley, F. C., M. Edwards, fig. 1.

Case and Column of Drawers.	Reference to McCoy's Synopsis: and Figures of Genera.	Names and References; Observations, &c.	Numbers and Localities.
		[**Palæocyclus præacutus,** Lonsdale, is the May Hill Sandstone species, see p. 84. I do not remember that it is found in Wenlock strata except in S. Wales.]	
F C		**Palæocyclus Fletcheri,** Milne Edwards (Brit. Fos. Cor. t. 57, fig. 3). A conical cup-shaped species, more like *Petraia* than any other. The buds take their rise from the surface of the cup.	a. 340, Dudley, F. C.
F C		**Palæocyclus rugosus,** Milne Edwards (Brit. Fos. Cor. p. 248, t. 57, fig. 4).	a. 341, Dudley, F. C.
G b 8	Pl. 1 B, fig. 18, p. 33.	**Clisiophyllum vortex,** McCoy, said to be from the Wenlock limestone—is without doubt the mountain limestone sp. *C. coniseptum,* Keyserling. Milne Edwards (Brit. Fos. Cor. t. 37, fig. 5). A common fossil in collections.	No doubt from the Carb. Limestone of Oswestry, figured specimens are a. 374.
		ECHINODERMATA.	
	Star-fishes.	**Lepidaster,** Forbes. Mem. Geol. Surv. Decade 3 (distinct from *Palæaster*).	
F C		**Lepidaster Grayii,** Forbes. Decade 3, pl. 1. Two fine large specimens.	a. 717, Dudley, F. C.; a. 716, Dudley, F. C., figured specimen.
	Crinoids or *Sea-lilies.*	Roots, and stems (often swelled by disease) are extremely common, so common as to characterize the whole formation. Separate plates of various genera are less common, but are found in the more shaly strata.	
	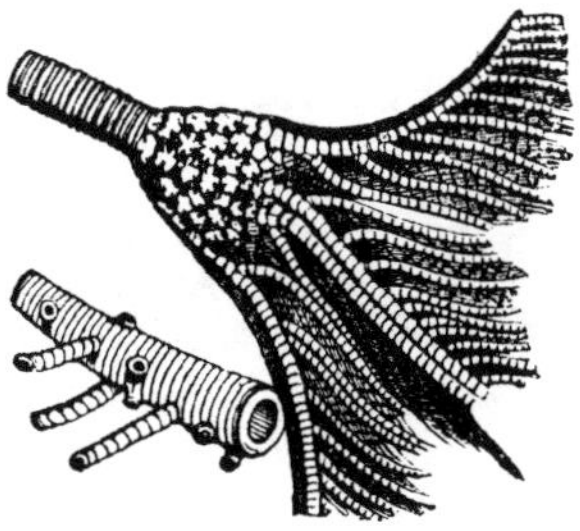	I adopt the order in which the genera are given in the Synopsis for convenience of reference. But it is not a natural order. Some genera have been since added; and many more remain to be described from the magnificent Fletcher Collection, and that of Mr John Gray of Hagley, now deposited in the British Museum. Mr C. Ketley of Smethwick has added largely to both collections, both cystideæ and crinoids.	

Case and Column of Drawers.	Reference to McCoy's Synopsis: and Figures of Genera.	Names and References; Observations, &c.	Numbers and Localities.
		a. Irregular Crinoids.	
		**Herpetocrinus,** n. g. (*Myelodactylus*, Hall. The name, based on false ideas, cannot be kept. The column is described as fingers and wrongly named, for the pierced column and finger-joints are common characters). I am constrained to form a new genus for about three species of crinoids which depart from all ordinary rules. A minute cup, all but obsolete, and with dichotomous narrow arms, branched like those of *Homocrinus*, Hall. This is set on a narrow cylindrical suberect stem, which soon however becomes compressed, then hollowed on one side, and bears close set auxiliary arms to the lowest part, so far as observed. The stems coiled up like an adder, resemble arms of crinoids rather than stems.	
F C		**Herpetocrinus Fletcheri,** n. sp. Stems two or three lines wide at most; with narrow, smooth, unridged rings, not tuberculate, arms repeatedly dichotomous, linear.	a. 384, Dudley, F. C.
		Other species exist in the Gray collection.	
		Cheirocrinus, n. g. (Austin named it *Pendulocrinus*, a bad name, for it is compounded of Latin and Greek). *Cheirocrinus* is now accepted generally. It was proposed in Siluria, 2nd ed. (commonly called 3rd ed.) 1859, by me.	
F C		**Cheirocrinus serialis,** Austin, sp. (Siluria, 2nd ed. Appendix, p. 535). The upper arm is greatly larger than the others, and the lower ones crowded under it.	a. 385, Dudley, F. C.; also cast of Gray's specimen (Brit. Mus.).
F C		**Cheirocrinus abdominalis,** n. sp. Shape like the last, but arms much more equal in size.	a. 386, Dudley, F. C.
F C		**Cheirocrinus gradatus,** n. sp. Arms nearly equal in size; simple, form not very much oblique.	a. 387, Dudley, F. C.

Case and Column of Drawers.	Reference to McCoy's Synopsis: and Figures of Genera.	Names and References; Observations, &c.	Numbers and Localities.
F C		**Cheirocrinus Fletcheri,** n. s. (*C.* sp. branched arm) Fletcher Coll. Siluria, 2nd ed. App. p. 535. The finest and most curious of all the species. It has repeatedly branched nodose arms, and is the largest of all, and will shew us the affinities with other genera at present obscure.	a. 388, Dudley, F. C.
		b. Crinoids with double rows of plates in the arms.	
	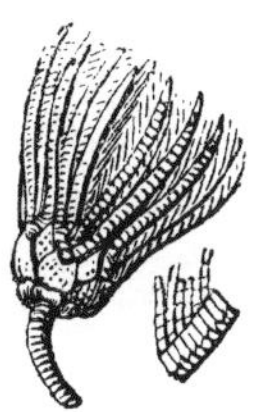	**Marsupiocrinus,** Phillips. The arrangement of the plates in the broad cup is that of *Eucalyptocrinus,* Goldfuss, and of *Hypanthocrinus.* But the proboscis is not gigantic and solid as in the latter genus: and we do not know enough of Goldfuss' figured genus.	
F C G d	p. 54.	**Marsupiocrinus cælatus,** Phill. (Siluria, 2nd ed. pl. 14, fig. 1, and p. 247, woodcut 55, figs. 1—3). A fine series, shewing young and old cups, arms, interior of arms, stomach surface retracted in rest, or produced into proboscis for feeding on the Gasteropoda. Mr John Gray found that *Acroculia Haliotis* was the favourite food. See woodcut in Siluria, Foss. 55, as above.	a. 667, with *Pseudocrinus;* a. 389, Dudley, F. C.; a. 390, feeding on *Acroculia,* Dudley, F. C.; a. 391, Dudley, interior of arms, Ketley Coll.; casts of Gray's specimens (Brit. Mus.).
		Syriocrinus, Hall. Very much resembles *Marsupiocrinus,* if it be not the same genus.	
	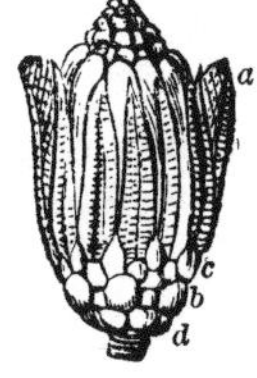 p. 57.	**Hypanthocrinus,** Phillips (Sil. Syst.), *Eucalyptocrinus,* Salter, &c. in Siluria, 2nd ed. There is no ground for so altering the generic name: though I myself was the first to propose it. The Devonian genus has a much shorter calyx, and several genera are now found to have the conspicuous star-like arrangement of plates round the interradials *b.*	

Case and Column of Drawers.	Reference to McCoy's Synopsis: and Figures of Genera.	Names and References; Observations, &c.	Numbers and Localities.
F C	*Eucalyptocrinus*, p. 58. 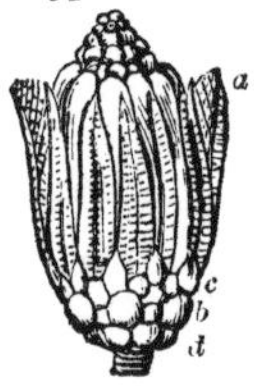	**Hypanthocrinus decorus,** Phill. (*Eucalyptocrinus*, Siluria, 2nd ed. pl. 14, fig. 2). The remarkable large *interbrachial* plates (*a*) placed between the arms readily distinguish this from any other genus. The convex plates of the cup *interradial* (*b*) plates, placed between the *radial* (*c*) plates which go directly to the arms from the cup or calyx (*basal plates, d*), distinguish it from *H. laciniatus*, and *H. granulosus*, both Dudley and Walsall species. The nomenclature of the plates of the cup of a crinoid is best given by De Koninck, in his monograph of the Belgian *Crinoidea* from the Carboniferous Limestone.	a. 392, Dudley, F. C.
		Hypanthocrinus granulatus, Lewis. London Geological Journal, 1847, p. 99, pl. 21. Geol. Mus. Jermyn St. has good Walsall ones.	
F C	*Eucalyptocrinus*, Pl. 1 D, fig. 2, p. 58.	**Hypanthocrinus polydactylus,** McCoy. An excellent species, easily distinguished from both the last by the many-branched arms. This species shews the affinity of *Hypanthocrinus* to *Marsupiocrinus* very easily.	a. 393, Dudley, F. C.; cast also from Gray's specimens, Brit. Mus.
	For figures of several of these genera, see *Siluria*, 3rd ed. pl. 13, figs. 4, 5.	**Dimerocrinus,** Phill. Sil. Syst. 1839. Still further removed than *Marsupiocrinus* from the last; there is yet much affinity. The interradial plates (*a*) are very conspicuous. The species very common at Dudley.	
F C G c		**Dimerocrinus icosidactylus,** Phill. (Siluria, 2nd ed. pl. 13, fig. 4). The twenty arms and three or four interradials distinguish this common form from the next species.	a. 670, Dudley, F. C. good, with *Glyptocrinus*. a. 669, ditto.
F C		**Dimerocrinus multiplex,** n. sp. Five or six or more interradials, none of which are conspicuously larger than the rest.	a. 404, Dudley, F. C.
F C		**Dimerocrinus uniformis,** n. sp. Cup all even, the sutures barely marking the smooth plates. Ten arms, which are parallel.	a. 405, Dudley, F. C.

Case and Column of Drawers.	Reference to McCoy's Synopsis: and Figures of Genera.	Names and References; Observations, &c.	Numbers and Localities.
F C		**Dimerocrinus decadactylus,** Phill. (Siluria, 2nd ed. pl. 13, fig. 5). Cup with convex highly ornamented plates. Ten arms which diverge. A common species. This grows large. A very fine specimen shews the arms and plates of the cup tuberculate.	a. 406, Dudley, F. C. a. 407, Dudley, F. C.
	p. 56. 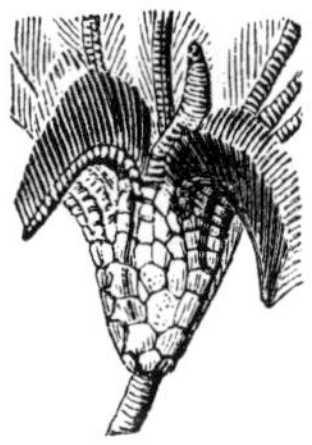	**Periechocrinus,** Austin. The largest and most common of Dudley crinoids, with a long conical cup—plated all the way up to the end of the bifurcation of the arms, which then, fifteen to thirty-four in number, are long, straight and unbranched; but of course not simple, being composed of a double row of bones. Proboscis massive, central. Stomach plates small. Stems moniliform, of thick joints. Range—May Hill Sandstone to Ludlow rock.	
F C	p. 56.	**Periechocrinus moniliformis,** Miller, spec. (Siluria, 2nd ed. pl. 13, figs. 1, 2). There are thirty-two or thirty-four arms. Pelvis conical, ridged by the radial plates. The arms branch and dichotomise.	a. 408, stem, a. 409, cup dissected, a. 410, arms all displayed, Dudley, F. C.; a. 411, proboscis (Ketley Coll.); a. 412, proboscis (Ketley Coll.); a. 417, Dudley, F. C.; a. 418, shews large pair of arms five - branched (Ketley Coll.).
F C	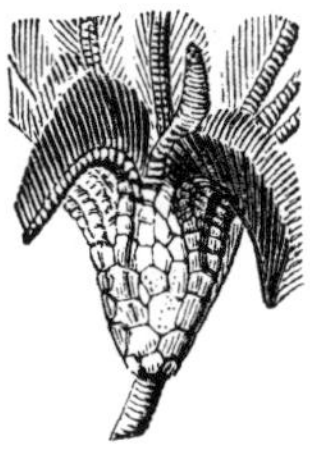	**Periechocrinus simplex,** n. s. Arms fifteen only. Cup smaller and broader, radials strongly ridged, plates radiated. A very distinct good species.	a. 413, Dudley, F. C.
F C		**Periechocrinus limonium,** n. s. Arms twenty-seven or twenty-eight, as in *moniliformis,* but cup ovate. Scarcely ridged by the radial lines. Plates thin, looks like *Hypanthocrinus granulosus.*	a. 414, Dudley, F. C.; a. 415, shews the stomach-plates; a. 416, young, Dudley, F. C.

Case and Column of Drawers.	Reference to McCoy's Synopsis: and Figures of Genera.	Names and References; Observations, &c.	Numbers and Localities.
		Mariacrinus, Hall. A good genus, closely allied to the above, but the two rows of plates in the arm separate early, and form distinct branches, which again branch, or are rather feathered by arms in great numbers. The cup is in five pieces.	
F C G d		**Mariacrinus flabellatus,** n. s. Named long ago by myself in the Fletcher Coll. at Dudley, and still unpublished.	a. 419, three large specimens and eight roots and stems, Dudley, F. C.; a. 662, good, Dudley, F. C.
		[**Actinocrinus,** proper, is sure to be found in the Wenlock limestone; since we have it from the Denbigh flagstone of Wales and Westmorland.]	
		Platycrinus, Miller. Some little doubt exists if this be truly a *Platycrinus,* since the arms are truly simple, not double-jointed. Still it is so like, that it is convenient; and the species does not belong to *Actinocrinus,* to which it was referred.	
F C G d G c 3	The figure in *Siluria* wants the stem.	**Platycrinus retiarius,** Phill. (Siluria, 2nd ed. pl. 14, fig. 9). Small cup, the first radials large and broad for the size of the creature. Plumes short.	a. 420, Dudley, F.C.; a. 665, good, Dudley, F. C.; a. 421, a large specimen shewing stem, Dudley, F. C; a. 422, shews pelvis well, Dudley, F. C.
F C		**Platycrinus pecten,** Salter, n. sp. A much larger cup, and long straight comb-like plumes to the fingers.	a. 423, Wren's Nest, F. C.
	p. 57.	**Glyptocrinus,** Hall? Very few plates and very short arm-plates.	
F C		**Glyptocrinus,** sp. 1. Like no other crinoid that I know.	a. 512, Dudley, F. C.
F C G c		**Glyptocrinus expansus,** Phill. (Siluria, 2nd ed. pl. 15, figs. 1, 2). Very doubtfully belonging to this genus. The cup is too large, and the series of plates not quite exact, as pointed out under *G. basalis,* in p. 45. It is convenient to leave it here for the present.	a. 508, interior of base of cup; a. 509; a. 510; a. 511, young with long stem, Dudley, F. C.; a. 669, good; a. 890, ditto; a. 889, uncrushed, round specimen, Dudley, F. C.

Case and Column of Drawers.	Reference to McCoy's Synopsis: and Figures of Genera.	Names and References; Observations, &c.	Numbers and Localities.
	 *C. rugosus*, p. 55.	**Crotalocrinus,** Aust. A most remarkable genus. In all other crinoids yet known the arms are free and the fingers free; in this they are all united by transverse processes into a flexible basket, just like wicker-work. To make the apparent anomaly greater, the arms *appear,* and appear only to start in great numbers from the edge of the globular calyx; whereas they really start from a point far inward on the stomach plates—the upper edge of the calyx being strongly and sharply inflected. Only one species is known in Britain. In Sweden another has been described by Müller, which really has the five fans of fingers distinct at the base, and overlapping each other. The affinity is with our next genus, *Cyathocrinus*, to many species of which there is a remote resemblance in the numerous fingers, and the structure of the cup is the same, viz. one set of subradial plates.	
F C G c	 p. 55.	**Crotalocrinus rugosus,** Miller, (Siluria, 2nd ed. Woodcut 55, fig. 4—7, p. 247, pl. 13, fig. 3).	a. 424, a. 425, a. 426, var., a. 427, a. 428, shew stomach-surface; a. 430, base of all the arms; a. 429, a. 432, a. 433, roots, Dudley, F. C.; a. 431*, Gray's cast; a. 671, the most perfect stomach-plates, balloon-shaped stomach (Ketley Coll.).
		Cyathocrinus, Miller. Cup of five basal pieces, with five intermediate (subradial) pieces, between which the arms originate.	
F C		**Cyathocrinus quinquangularis,** Phill. (not Mill.), *Rhodocrinus,* Sil. Syst. t. 18, fig. 5.	a. 435, Dudley, F. C.
F C		**Cyathocrinus** (sp. 1). Like *C. goniodactylus,* with similar cup, and general structure, but greatly thicker arms.	a. 499, Dudley, F. C.
F C		**Cyathocrinus** (sp. 2) **decadactylus,** Salter. Ten single arms, unbranched, and with very large thick tentacles.	a. 494, Dudley, F. C.

Case and Column of Drawers.	Reference to McCoy's Synopsis: and Figures of Genera.	Names and References; Observations, &c.	Numbers and Localities.
F C		**Cyathocrinus** (sp. 3) **quindecimalis,** Salter, MSS. Fifteen arms, like those of the last species, but with even thicker tentacles. N.B. These last two species form a very peculiar group—unlike other Wenlock species.	a. 495, Dudley, F. C.
G c 2		**Cyathocrinus** (sp. 4) **squamiferus,** MSS. With very broad scale-like joints to the arms, which seem also thin and flat.	a. 452, Dudley, F. C.
G c 2		**Cyathocrinus** (sp. 5). Small species, like *C. arthriticus*, but with arm-plates much longer in proportion and of course narrower.	a. 486, Dudley, F. C.
F C		**Cyathocrinus arthriticus,** Phill. (Siluria, pl. 14, fig. 7). A species with greatly extended upper angles to the joints.	a. 447, young; a. 448, a. 449, Dudley, F. C.
G d		**Cyathocrinus goniodactylus,** Phill. (Siluria, 2nd ed. pl. 14, fig. 3). The commonest species—of all ages; the specimens often appear to form different species.	a. 663, a. 664, good, Dudley, F. C.
G c 2		**Cyathocrinus,** 2 new sp A good deal like the young *C. goniodactylus*, but with thinner and more dichotomous arms.	a. 444, a. 445, Dudley, F. C.
G c 2		**Cyathocrinus,** sp. Cup of an allied species (roughly reticulate).	a. 446, Dudley, F. C.
F C		**Cyathocrinus,** sp. A similar cup.	a. 450, Dudley, F. C.
F C		**Cyathocrinus,** sp. With subangular arms like the two next, but distinct.	a. 496, Dudley, F. C.
G c 2		**Cyathocrinus** (sp. 6) **monile,** MSS. Rounded and much-branched arms.	a. 487, Dudley, F. C.
F C		**Cyathocrinus nodulosus,** MSS. Much like last, but only half the number of arms.	a. 487*, Dudley, (Ketley Coll.).
F C G d		**Cyathocrinus capillaris,** Phill. (Siluria, 2nd ed. pl. 15, fig. 3). A pretty species, with slender, long and much-branched arms, and a very much ornamented calyx.	a. 488, a. 489, a. 490, Dudley, F. C.; a. 491, (Ketley Coll.); a. 668, good, Dudley, F. C.

Case and Column of Drawers.	Reference to McCoy's Synopsis: and Figures of Genera.	Names and References; Observations, &c.	Numbers and Localities.
G c 2		**Cyathocrinus** (sp. 7) **scoparius,** MSS. s. Much like the last, but the arms more subdivided, and even apparently tufted; and the terminal stomach plates appear to have also ended in brush-like tufts.	a. 491, Dudley, F. C.
F C		**Cyathocrinus punctatus** (*Apiocrinus punctatus,* Hisinger, Leth. Suec. p. 89, t. 25, fig. 2).	a. 436, a. 437, Dudley, F. C.
G c 2		**Cyathocrinus** (sp. 8) **Ichthyocrinoides.** Like young *punctatus,* but with very short close arm-joints, and (smooth? pelvis).	a. 492, Dudley, F. C.
G c 2		**Cyathocrinus** (sp. 9) **arboreus,** MSS. Much branched, with round arms and fingers, and minute pelvis.	a. 493, Dudley, F. C.
F C		**Cyathocrinus** (sp. 11). Like *arboreus,* but greatly branched above.	a. 497, Dudley, F. C.
		Taxocrinus, D'Orbigny. Distinguished from *Cyathocrinus,* which it much resembles, by the want of *any subradial plates;* the arms are simple, not double; and there are in general only a few interradial plates (*a*) in the cup. On the anal side only are those interradials brought down to the cup.	
F C	p. 53.	**Taxocrinus tuberculatus,** Miller (Siluria, 2nd ed. pl. 14, figs. 5, 6). The commonest species of the genus, easily discriminated by the strong tubercles all along the plates and arm-joints. Specimens of all ages shew that young crinoids have much fewer arms, or rather fingers, than old ones.	a. 500, a. 501, Dudley, F. C.
G c 3		**Taxocrinus tesseracontadactylus,** Hisinger, sp. (Leth. Suecica, t. 25, fig. 4). It is very doubtful if the *Taxocrinus simplex* of Phillips be not this Swedish species.	a. 500*, Dudley, F. C.
F C		**Taxocrinus simplex,** Phill. (Sil. Syst. pl. 18, fig. 8).	a. 502, a. 503, a. 504, Dudley, F. C.; a. 506, interior (Ketley Coll.).
F C		**Taxocrinus marmoratus,** Salter, n. s. The plates of the cup and the basal portions of the arms roughly tubercular.	a. 505, a. 507, Dudley, F. C.

Case and Column of Drawers.	Reference to McCoy's Synopsis: and Figures of Genera.	Names and References; Observations, &c.	Numbers and Localities.
F C		**Taxocrinus granulatus,** Salter, n. s. (Like *marmoratus.*)	a. 530, Dudley, F. C.
F C		**Taxocrinus nanus,** Salter, n. s. Its squat form and single interradial easily introduce us to the next genus.	a. 527, Dudley, F. C.
		Ichthyocrinus, D'Orbigny. Distinguished from *Taxocrinus* solely by the want of any interradial plates; the arms lie close together, and start at once from the basal plates or pelvis.	
F C	p. 54.	**Ichthyocrinus pyriformis,** Phill. (Siluria, 2nd ed. pl. 14, fig. 8). A most elegant crinoid, in which the short joints of the arms and fingers range equally across the whole figure, which is pyriform, and apparently never spread out.	a. 498, a. 529, Dudley, F. C.
F C		**Ichthyocrinus Bacchus,** MSS. n. s. The depressed short form and very short plates so much resemble those of *Taxocrinus nanus,* that some close observation is needful to distinguish this, which has no interradial plates, from the former, which possesses very narrow ones. (J. W. S.)	a. 528, Dudley, F. C.
		CYSTIDEÆ, Von Buch.	
	Globe crinoids or *Globe-lilies.*	Crinoids in general plan: these are almost deprived of the characteristic arms; a few tentacles round a wide stomachal surface (*ambulacral,* Huxley), the aperture of which is not yet known, are in a few genera exchanged for simple recurved arms (*Pseudocrinites, Apiocystites,* &c., and other Upper Silurian genera): these are laid in special grooves, so as to be as much out of the way as possible. The whole group has relations with *Pentremites,* through which it seems to approach the true *Echinei,* &c. The anal, usually called ovarian opening, is covered by projecting plates. (Billings.)	

Case and Column of Drawers.	Reference to McCoy's Synopsis: and Figures of Genera.	Names and References; Observations, &c.	Numbers and Localities.
F C	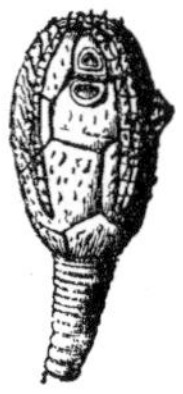	**Apiocystites pentremitoides,** Forbes (Mem. Geol. Surv. Vol. II. Pt. 2, pl. 15, figs. 1—9), Siluria, 2nd ed. Woodcut 54, fig. 4.	a. 588, Dudley, F. C.; a. 589, a. 590, Dudley, (Ketley Coll.); figured by Forbes.
		Prunocystites, Forbes (Mem. Geol. Surv. Vol. II. Pt. 2, p. 503, pl. 16).	
F C		**Prunocystites Fletcheri,** Forbes (ib. pl. 16, figs. 1—4, Siluria, 2nd ed. Foss. 54, fig. 3).	a. 586, a. 587, Dudley, F.C.
		Pseudocrinites, Pearce. The first noticed of this group in Britain; but of course supposed to be a spurious kind of crinoid, with less than five arms. The three pectinated rhombs are thick-edged.	
F C G d		**Pseudocrinites quadrifasciatus,** Pearce (Mem. Geol. Surv. ib. pl. 13, fig. 1). The common species in Dudley limestone; rarely found elsewhere (Siluria, 2nd ed. Foss. 54, fig. 2).	a. 594, stem; a. 595, tentacles and rhombs; a. 596, young; a. 597, grooves for arms; a. 666, good, Dudley, F. C.
F C		**Pseudocrinites bifasciatus,** Pearce. (Forbes, ib. p. 496, pl. 11). Only two arms, arranged as in *P. magnificus.*	a. 591, a. 592, a. 593, figured specimens, Dudley, F. C.
F C		**Pseudocrinites magnificus,** Forbes (ib. pl. 12, fig. 1, Siluria, 2nd ed. Foss. 54, fig. 1). A very large species, the largest body of any (in Britain) with rhombs.	a. 598, Cast (Gray Coll. B. Mus. *cast*), Dudley, F. C.
F C		**Pseudocrinites oblongus,** Forbes (ib. pl. 14, figs. 1—14, p. 499).	a. 599, figured, Dudley, F.C.
		**Echino-encrinites,** Volborth. A transverse long slit marks the place of the mouth; some doubt if this be quite the same genus as the Russian Upper Cambrian fossil described by Volborth: still more if *E. baccatus* belong to the same genus.	

Case and Column of Drawers.	Reference to McCoy's Synopsis: and Figures of Genera.	Names and References; Observations, &c.	Numbers and Localities.
F C		**Echino-encrinites armatus,** Forbes (Mem. Geol. Surv. ib. pl. 18, figs. 1—12). A compressed square species, granulated all over, a boss on each plate (Siluria, 2nd ed. Foss. 54, fig. 6).	a. 582, a. 583, Wenlock Shale, Walsall, figured by Forbes; a. 584, a. 585, ditto, Malvern, figured edge of mouth, and shews the rhombs.
F C		**Echino-encrinites baccatus** (Mem. Geol. Surv. II. Pt. 2, pl. 17, figs. 1—10), Siluria, 2nd ed. Foss. 54, fig. 5).	a. 606, figured; a. 607, a. 608, Dudley, F. C.
		Ateleocystites, Billings. Founded on a Canadian fossil from the Trenton limestone of Ottawa. The flattened form and single anal plate easily distinguish it.	
F C	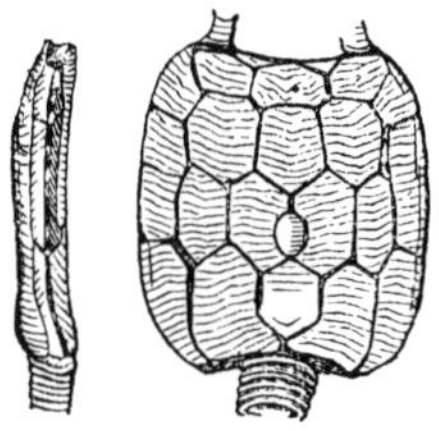	**Ateleocystites Fletcheri,** Salter, n. s. Much larger and broader than the Canadian species, rare, and yet not quite local.	a. 600, ovarian side; a. 601, ventral; a. 602, dorsal side; a. 603, ditto; a. 604, young; a. 605, base and part of stem, Dudley, F. C.
		Pisocrinus, De Koninck. The globular calyx, with apparently no rhombs, and long tentacular arms (Dr Grindrod of Malvern has them in his cabinet).	
F C		**Pisocrinus pilula,** De Koninck, Bull. Acad. Roy. des Scien. de Bruxelles, Vol. IV. 2 ser. p. 106, t. 2, figs. 8—11.	b. 450, Wenlock Shale, Malvern.
F C		**Pisocrinus ornatus,** id. 2 ser. Vol. IV. p. 107.	b. 452, Dudley, F. C.
		ANNELIDA.	
F C	Worm-tracks, tubes and shells.	**Cliona?** or **Vioa.** Borings of a minute annelide inside the shell of *Orthoceras.*	a. 467, Dudley, F. C.
F C	p. 63.	**Cornulites serpularius,** Schloth. (Siluria, 2nd ed. pl. 16, figs. 3—10). A shelly tube with cellular varices (knots) like *Tentaculites,* also grows in knots of 3, 4—8 young shells, separating afterwards.	a. 531, variety with very close rings in youth; a. 686, a var. with very remote rings in youth, Dudley, F.C.
F C G c 4	do. 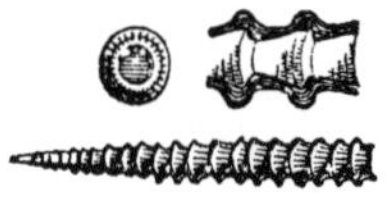	**Tentaculites ornatus,** Sow. (Siluria, 2nd ed. pl. 16, fig. 11). In Münster's coll. this is called *T. striato-costatus.* It has a wide range.	a. 688, (Münster's), a. 688*, Dudley, F. C.

Case and Column of Drawers.	Reference to McCoy's Synopsis: and Figures of Genera.	Names and References; Observations, &c.	Numbers and Localities.
F C G c 4 G c 5	*Serpulites dispar,* p. 132.	**Scolecoderma,** Salter. The tubes of mud-loving annelides are thin and membranous usually. (*S. dispar,* Salter, is a true *Serpulites.*)	a. 897, a. 695, Dudley, F. C.; a. 698, Dinas Bran, in Wenlock Shale.
		[**Turrilepas,** H. Woodward. Shelly plates arranged in four rows, unsymmetrical, carinated, and imbricated, with a uniformly sculptured surface, destitute of lateral processes. Considered to be related to *Loricula, Scalpellum,* and other pedunculated cirripeds. Mr Salter considered it an Annelide tube, shelly, of numerous plates imbricating backwards, and proposed the name *Oploscolex* for it. *Editor.*]	
F C G c 4		**Turrilepas Wrightianus,** De Kon. (*Chitcn,* De Kon. Bull. Acad. Roy. Bruxelles, 2 ser. Vol. III. p. 199, tab. 1, fig. 2). *Turrilepas,* Woodward (Quart. Journ. Geol. Soc. Vol. 21, p. 486, pl. 14, fig. 1).	a. 532, Dudley, F. C.; b. 729, Dudley (Mr Ketley).
G c 4		**Turrilepas Ketleyanus,** Salter, sp. MS.	b. 730, Dudley (Mr Ketley).
		CRUSTACEA Phyllopoda.	
	Bivalve Shrimps, Water-Fleas, and Trilobites.	Though the Upper Cambrian rocks are the metropolis of the Trilobite group, there are a good many genera (all, with two exceptions, the same as Cambrian) and the species and individuals are very numerous in Wenlock and Dudley strata. Small Phyllopods and large shrimp-like forms (*Ceratiocaris*) begin to be abundant and numerous in species.	
G c 5	p. 136.	**Ceratiocaris,** sp. The genus will be more fully described in the Ludlow rocks, and it is doubtful if this one be of Wenlock age.	a. 694, Dudley, in Upper Shale.
G d	Pl. 1 E, fig. 2, p. 135.	**Beyrichia Kloedeni,** McCoy. (Ann. N. Hist. 2 ser. Vol. XVI. pl. 6, figs. 7—9). The genus is a most natural and convenient one for those small-lobed and tubercled bivalve crustacea which it represents.	a. 886, McCoy's figured specimen, Myddfai, Llandovery; b. 731, Llanfair, Welchpool.

Case and Column of Drawers.	Reference to McCoy's Synopsis: and Figures of Genera.	Names and References; Observations, &c.	Numbers and Localities.
		CRUSTACEA Trilobita.	
	p. 150.	**Lichas,** Dalm. Almost as curiously constructed as *Acidaspis,* and apparently a close ally. The lobes of the glabella are reversed, as to size, from the usual plan.	
F C		**Lichas scutalis,** Salter MSS. (*L. verrucosus,* Salter, Mem. Geol. Surv. Vol. II. Pt. 1, pl. 8, fig. 7, not Eichwald).	a. 954, Malvern, Wenlock Shale, Dudley, F. C.
F C		**Lichas Salteri,** Fletcher (Quart. Journ. Geol. Soc. Vol. VI. t. 27, fig. 9, p. 237).	a. 953, Dudley, F. C.
F C G c 6		**Lichas Grayii,** Fletcher and Salter (ib. Vol. VI. p. 237, t. 27, fig. 8).	b. 26, b. 27, figured specimens, a. 699, Dudley, F. C.
F C		**Lichas hirsutus,** Fletcher and Salter (ib. Vol. VI. p. 236, t. 27, figs. 6, 7).	a. 961—a. 965, Dudley, F. C.
F C		**Lichas Barrandii,** Fletcher and Salter (ib. Vol. VI. p. 238, t. 27, fig. 10. Siluria, 2nd ed. Foss. 63, fig. 3).	b. 28, b. 29, figured, Dudley, F. C.
F C G d	*Acanthopyge,* p. 151.	**Lichas anglicus,** Beyrich, sp. (ib. t. 27, figs. 1—5, Siluria, 2nd ed. Foss. 63, fig. 1). A most curious glabella, with puffed lobes like the cheeks of a baboon.	a. 958, a. 959, a. 960, good; a. 955, figured, a. 956, a. 957, Dudley, F.C.; b. 732, Dudley.
		Cheirurus, Beyrich. A very natural group, allied to *Phacops,* and yet with affinities for *Encrinurus, Cybele,* &c. The tail is palmate; the body rings strongly sculptured.	
F C	*Ceraurus Williamsi,* Pl. 1 F, fig. 13, p. 155.	**Cheirurus bimucronatus,** Murch. (Salter, Mon. Brit. Tril. t. 5, figs. 1—5, Siluria, 2nd ed. Foss. 64, fig. 4, pl. 3, fig. 5, pl. 19, figs. 10, 11).	a. 533, a. 534, figured specimens (Salter) in Decade 7, Geol. Surv. pl. 2, figs. 4—6, Dudley, F. C.
		Encrinurus, Emmrich. Clavate glabella, pedunculate eyes, and plain sub-cylindrical pointed body rings; tail of many joints.	

Case and Column of Drawers.	Reference to McCoy's Synopsis: and Figures of Genera.	Names and References; Observations, &c.	Numbers and Localities.
F C G c 4	p. 158.	**Encrinurus punctatus,** Brünnich (Decade 7, Geol. Surv. t. 4, fig. 15, Siluria, 2nd ed. Foss. 14, fig. 10; 64, fig. 5, pl. 10, fig. 5).	a. 539, figured specimen, Fletcher, Geol. Journ. Vol. VI.; a. 541, a. 542, labrum; a. 540 shews eyes well; Dudley, F. C.; b. 732, Myddleton Park, Caermarthen.
F C G d	*Zethus,* p. 157.	**Encrinurus variolaris,** Brong. (ib. 7, t. 4, figs. 13, 14, Siluria, 2nd ed. pl. 18, fig. 9, Foss. 64, fig. 6). Tail blunt, of few rings.	a. 537, fine rolled-up specimen; a. 538 shews the labrum in position; Dudley, F. C.
	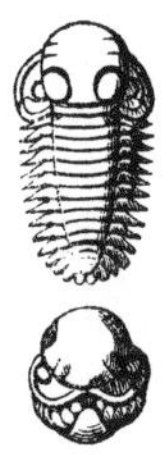	**Sphærexochus,** Beyrich. A most extreme form of *Cheirurus,* in which all the sculpture of the body rings is lost.	
F C		**Sphærexochus mirus,** Beyrich (Salter, Mon. Brit. Tril. t. 7, figs. 1—6, and Decade 7, pl. 7, Siluria, 2nd ed. Foss. 64, fig. 1).	a. 543, tail figured, Decade 7, Geol. Surv., Dudley, F. C.
	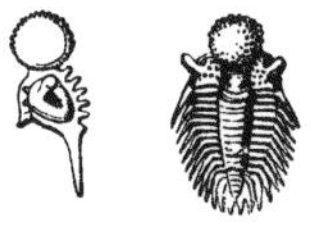	**Staurocephalus,** Barr. An exaggerated *Cheirurus* with a strongly clavate and pedunculated glabella. Body rings 10, tail pectinate.	
F C	Pl. 1 F, fig. 15, p. 153.	**Staurocephalus Murchisoni,** Barr. (Decade 11, Geol. Surv. pl. 5, figs. 1—4, Salter, Mon. Brit. Tril. p. 84, pl. 7, figs. 13—20).	a. 544, Dudley, F. C.
F C G d	p. 160. 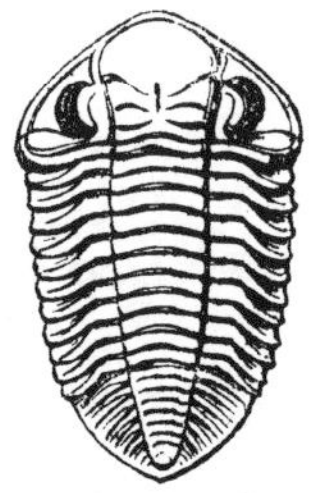	**Phacops Downingiæ,** Murch. (Salter, Mon. Brit. Tril. pl. 2, figs. 17—25, Siluria, 2nd ed. Foss. 64, fig. 3, pl. 18, figs. 2, 5).	a. 568, to a. 573, and a. 575, a. 576, Dudley, F.C.; a. 888, fine specimen, Dudley, F. C.
F C	As *Odontochile caudata,* p. 160.	**Phacops Downingiæ,** var. **constrictus,** Salter (Mon. Brit. Tril. p. 27, pl. 2, figs. 13—16).	a. 574, Dudley, F. C.
F C G d	*Portlockia,* p. 163.	**Phacops Stokesii,** M. Edw. (Salter, Mon. Brit. Tril. p. 21, pl. 2, fig. 6, Siluria, 2nd ed. pl. 10, fig. 6).	a. 578, Wenlock Shale, Dudley, F. C.; a. 895, Llangynyw Rectory, Welchpool; a. 887, Dudley, F. C.
F C		**Phacops Musheni,** Salter (Mon. Brit. Tril. p. 23, pl. 2, figs. 7—12).	a. 577, rolled-up specimens, a. 579, Dudley, F. C.

Case and Column of Drawers.	Reference to McCoy's Synopsis : and Figures of Genera.	Names and References ; Observations, &c.	Numbers and Localities.
F C G c 4 G d	*Odontochile,* p. 160.	**Phacops caudatus,** Brongn. (Mon. Brit. Tril. pl. 3, figs. 4—18, pl. 4, figs. 1—5. Siluria, 2nd ed. pl. 17, fig. 2, pl. 18, fig. 1).	a. 557, a. 557*, large, a. 558, a. 559, interior, two specimens; a. 560, large eyes; a. 561, tail; a. 562, young; a. 563, labrum; a. 564, labrum, Dudley, F. C.; a. 654, interior, Dudley; a. 691, cast of Mr Lightbody's specimen from Ledbury; a. 692, the specimen from near Llandeilo, wrongly said by the collector to be from Lower Bala.
F C G c 4	*Odontochile,* p. 161.	**Phacops longicaudatus,** Murch. (Mon. Brit. Tril. pl. 3, figs. 19—28, Siluria, 2nd ed. pl. 17, figs. 3—6).	b. 735, W. Shale, Myddleton Park; a. 567, Malvern, F. C.
F C		**Phacops longicaudatus,** var. β **Grindrodianus** (ib. pl. 3, figs. 2—28).	a. 565, a. 566, Wenlock Shale, Malvern, F. C.
G c 4		**Phacops longicaudatus,** var. **armiger,** Salter (ib. pl. 2, figs. 19—21).	b. 736, b. 737, Burrington, Cheney Longville.
G c 5		**Phacops tuberculato-caudatus,** Salter (ib. pl. 4, fig. 1), probably var. of *P. caudatus.*	a. 696, Lower Ludlow, Dudley, F. C.
		Calymene, Brongniart. The most compact and elegant form of the whole Trilobite group; moderate head and glabella; thirteen body rings and tail of six or seven joints; labrum notched.	
F C G c 4 G d	p. 165. As *C. subdiademata,* Pl. 1 F, figs. 9, 10, p. 166.	**Calymene Blumenbachii,** Brongn. (Salter, Mon. Brit. Tril. pl. 8, figs. 7—16, Siluria, 2nd ed. pl. 17, fig. 1, pl. 18, fig. 10). The young state is known in Sweden as *C. pulchella.* The older form is figured under all sizes, shapes and names in geological works. It is the Dudley locust of collectors.	a. 547, young; a. 548, older; a. 549, young; a. 550, young; a. 551, a. 552, good interiors; a. 553, largest known, except Mr Mathew's specimen (of which b. 733 is a cast); a. 554, expanded specimen; a. 655, fine large specimen; a. 656, a. 657, good specimens, Dudley, F. C.; a. 690, Coed Sion, Llangadoc.

Case and Column of Drawers.	Reference to McCoy's Synopsis: and Figures of Genera.	Names and References; Observations, &c.	Numbers and Localities.
G d	p. 167.	**Calymene tuberculosa,** Salter (Decade 2, pl. 8, figs. 1—8, Siluria, 2nd ed. 18, fig. 11). Distinguished by the narrow axis and horizontal pleuræ, also the produced front.	b. 734, Wenlock Shale, Burrington, Cheney Longville.
	p. 167. 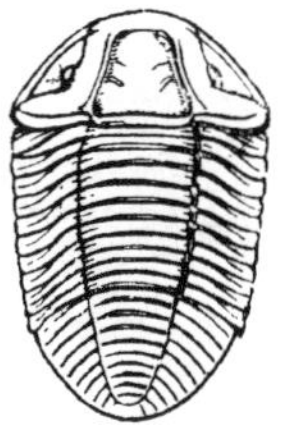	**Homalonotus,** König. A large Trilobite very closely related to *Calymene*, but with the axis so broad and so little raised above the pleuræ as to make the back even (ὁμαλός).	
F C G c 4 G c 6		**Homalonotus delphinocephalus,** Green (Sil. Syst. t. 7 *bis*, fig. 1, Siluria, Foss. 18, fig. 1, Salter, Mon. Brit. Tril. pl. 11, figs. 1—11). Abundant in Dudley limestone. The tail has a very short mucro.	b. 738, Falfield, Tortworth (Earl Ducie); a. 555, Dudley, presented by Captain Fletcher; a. 556, cast of Blackwell's specimen, Geol. Surv. Coll.
G c 4	*H. Knightii*, p. 168.	**Homalonotus Johannis,** Salter (Mon. Brit. Tril. pl. 12, fig. 11, pl. 13, figs. 1—7).	a. 693, Myddleton Park, Caermarthen.
F C G d		**Illænus insignis,** Hall (Salter, Mon. Brit. Tril. t. 27, figs. 6, 7. *Ill. Barriensis*, Sil. Syst. and Siluria, 2nd ed. pl. 17, figs. 9—11). Differs by gibbous head, position of eyes, and all characters from the common Woolhope species, *I. Barriensis*, Murch.	a. 580, Dudley, F. C.; a. 581 (rostral shield of do.); a. 653, good, Dudley, F. C.
	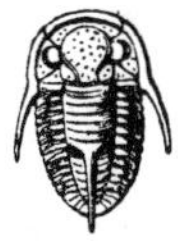	**Cyphaspis,** Burmeister. A genus of small trilobites related to *Proetus*, but with basal glabella lobes strongly circumscribed, and fourteen or fifteen body rings, rarely eleven or twelve.	
F C G c 4	*Harpidella megalops*, p. 143.	**Cyphaspis megalops,** McCoy. (Decade 7, Geol. Surv. pl. 5, Siluria, 2nd ed. Foss. 64, fig. 2).	a. 545, a. 546, a. 547, Dudley, F. C.; a. 689, Dudley, described by McCoy.
	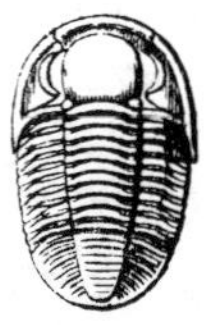	**Proetus,** Steininger (Mem. Geol. Soc. France). Allied to the terminal forms (Carboniferous era) of the *Asaphus* group of Trilobites. It has a rostral shield like *Phillipsia* and *Cyphaspis*, and the species vary very much in outline and number of body rings, eight to twelve (*Barrande*); a. 700 is one of the undescribed forms.	
G c 6	*Forbesia*, p. 174.	**Proetus,** spp. Two or three undescribed forms from Dudley and Wenlock, and Tortworth (Falfield) and Ledbury.	a. 700, Dudley, F. C.

Case and Column of Drawers.	Reference to McCoy's Synopsis: and Figures of Genera.	Names and References; Observations, &c.	Numbers and Localities.
	Forbesia, p. 174.	**Proetus latifrons,** McCoy. (Mem. Geol. Surv. Vol. II. pt. 1, t. 6, fig. 1.)	
F C		**Proetus Fletcheri,** Salter. A broader species in all parts, more like *P. Ryckholti,* than *P. latifrons,* McCoy.	a. 825, good, a. 828, Dudley, F. C.
F C	*Forbesia,* p. 174.	**Proetus Stokesii,** Murch. (Siluria, 2nd ed. pl. 17, fig. 7). Related to the *Cyphaspis* by its small glabella and strong basal lobes.	a. 899, a. 900, good, long head spines, a. 901, a. 902, interior, Dudley, F. C.
	p. 152.	**Acidaspis,** Murchison (*Odontopleura,* Emmrich, a little later in the same year). A curious genus allied to *Lichas.* The lobes of the glabella are scarcely separated from the cheeks. Spines ornament every part, differing according to species.	
F C G c 5		**Acidaspis coronatus,** Salter (Quart. Journ. Geol. Soc. Vol. XIII. p. 210). The free cheeks have very thick sinuous head-angles.	a. 658, a. 697, Dudley; b. 30, to b. 33, good (figured by Salter), probably from the Lower Ludlow Shale, Dudley, F. C.
F C	p. 152.	**Acidaspis Brightii,** Murch. (Siluria, 2nd ed. Foss. 64, fig. 8, pl. 18, figs. 7, 8).	a. 974, a. 975, figured in Fletcher; a. 976, young, good; Dudley, F. C.
		Acidaspis dama, Fletcher and Salter (Morris, Catal. Brit. Foss. 2nd ed. p. 99).	
F C		**Acidaspis crenatus,** Emmrich (Neus Jahrb. 1845, p. 44. *Ceraurus,* Löven, Ofvers Kon. Vet. Ak. 1845, t. 1, fig. 1, Siluria, 2nd ed. 1859, Appendix, p. 537).	a. 966, good; Dudley, F. C. a. 967, cast (Gray Coll.).
		Acidaspis dumetosus, Fletcher and Salter (Morris, Catal. p. 99).	
F C		**Acidaspis quinquespinosus,** Fletcher and Salter (Morris, Catal. p. 99).	a. 969 (with a. 968); a. 970, cast, Gray Coll.; a. 971, a. 972, figured; a. 973, seven spines; Dudley, F. C.
F C		**Acidaspis Barrandii,** Fletcher and Salter (Morris, Catal. 2nd ed. p. 99, Siluria, 2nd ed. Foss. 64, fig. 9). This belongs to *Trapelocera* of Corda, a good sectional name (Siluria, 2nd ed. Foss. 64, fig. 9).	a. 968, tail, in slab with *Acidaspis quinquespinosus;* a. 970*—973*, figured in Fletcher and Salter; Dudley, F. C.

Case and Column of Drawers.	Reference to McCoy's Synopsis: and Figures of Genera.	Names and References; Observations, &c.	Numbers and Localities.
		BRACHIOPODA (Palliobranchiata). Are easily divisible into (1) horny and hingeless shells (*Lingula, Discina, Obolus*), and (2) hinged calcareous shells, *Orthis, Spirifer,* &c. The passage from the young Lamellibranch to the *Brachiopod* is easily seen by laying open a young *Anodon* (Rathke's Memoir). The heart is double, the intestine duplicate; the form of the opened valves, like that of *Orthis biloba,* and the median plate in so many Brachiopods, is the sure indication of the real nature of these otherwise anomalous Bivalves.	
		Crania, Retzius. A well-known recent genus, which, with slight modifications, and very slight ones, has persisted from the Lowest Bala group to the present day. It inhabits deep water usually.	
F C	*Orbiculoidea,* p. 189.	**Crania implicata,** Sow. sp. (*Patella?* in the Silurian System. *Orbiculoidea* of D'Orbigny. *Crania,* Salter, Siluria, 2nd ed. pl. 20, fig. 4, Davidson, Sil. Brach. p. 80, pl. 8, figs. 13—17). The shell is a minute oval, imbricate outside, and with strong muscular scars.	a. 395, Dudley, F. C.
G c 7 F C		**Crania Grayii,** Davidson. (Sil. Brach. pl. 8, figs. 22—24.)	a. 713, Dudley, F. C., three specimens.
F C		**Trematis (*Discina*) Siluriana,** Davidson. (Sil. Brach. pl. 8, figs. 19—20.)	a. 809, a. 810, Dudley, F. C.
	Pl. 1 H, figs. 4, 5, p. 255. See *Athyris.*	**Spondylobolus,** McCoy. A genus unfortunately founded in mistake; a species of *Meristella* (probably *M. obovata*) being so pressed in shale, as to thrust the teeth through the opposite valve, and give rise to deceptive appearances. The formation also (this was not the Professor's error) is erroneous. The shell comes not from the black Lower Bala rocks of Builth, but from the equally black Wenlock shale. See No. 382.	

Case and Column of Drawers.	Reference to McCoy's Synopsis: and Figures of Genera.	Names and References; Observations, &c.	Numbers and Localities.
	p. 188.	**Siphonotreta anglica,** Morris (Ann. Nat. Hist. 2 ser. Vol. 4, tab. 7, fig. 1. Davidson, Sil. Brach. p. 75, t. 8, fig. 1. Siluria, 3rd ed. Foss. 58, fig. 10). [The genus *Siphonotreta* was established in 1845 by M. de Verneuil, for certain Brachiopod shells having well-defined characters which separate it from *Crania* and *Terebratula.* There are two British species. *Editor.*]	Dudley.
	p. 190.	**Discina** (*Orbicula* of old writers). Also a very old genus indeed. Under slight modifications it persists from Cambrian, even Lower Cambrian times, to the present day. And like *Crania* is a deep water form. The valves slide one on the other by the action of the muscles, and the byssus comes out of a subcentral foramen in the lower (ventral) valve.	
G c 7		**Discina Verneuilii,** Davidson (Sil. Brach. pl. 6, fig. 5).	a. 712, Dudley, F. C.; an inch long. Good specimen in Sharpe's cabinet, Geol. Society.
G c 7		**Discina striata,** Dav. (Sil. Brach. p. 191, pl. 6, figs. 1—4, Siluria, pl. 20, fig. 3).	a. 711, Dudley, F. C., Lower Ludlow ?
G c 7	p. 190.	**Discina Morrisii,** Davidson (Sil. Brach. pl. 7, figs. 10—12). It differs in its smooth shining surface from the *D. Forbesii.*	a. 718*, Dudley, F. C.
F C G c 7		**Discina Forbesii,** Davidson (Siluria, Foss. 57, fig. 11, p. 250, Mem. Geol. Surv. Vol. II. pt. 1, p. 371, pl. 26, fig. 2. Davidson, Sil. Brach. pl. 7, figs. 14—18). Both valves equally or nearly equally convex, forming the genus *Orbiculoidea,* of D'Orbigny.	a. 718, a. 719, Dudley, F.C.; a. 383, Keeper's Lodge, Golden Grove, W. Shale.
		Discina, sp.	
	See p. 250.	**Lingula,** Bruguière. The earliest known, and most common of all the Brachiopod group. Its structure has been described well by Dr Woodward in his Manual, and very wrongly referred by Prof. McCoy to the	

Case and Column of Drawers.	Reference to McCoy's Synopsis: and Figures of Genera.	Names and References; Observations, &c.	Numbers and Localities.
		Lamellibranchs. All Brachiopods differ from Lamellibranchs or ordinary bivalves, by the development of a ventral shell, and the absence of a foot. The hinge is different, no cartilage supporting it, while the opening of the valves is effected by muscles, as Prof. McCoy first shewed (loc. cit.). Prof. Owen has dissected the animal (Phil. Trans. and Davidson's Introductory Memoir to the British Fossil Brachiopoda).	
G c 7	p. 253.	**Lingula Lewisii,** Sow. (Davidson, Sil. Brach. p. 35, t. 3, figs. 1—6). A broad rectangular species, the front rounded. Siluria, pl. 20, fig. 5.	a. 708, Dudley Shales (Lower Ludlow ?), F. C.
G c 7 G d 1 F C		**Lingula Symondsii,** Salter (Davidson, Sil. Brach. p. 45, t. 3, figs. 7—17).	a. 706*, Cheney Longville (W. Shale); a. 710, a. 707, a. 706, Dudley, Wenlock Shale, F. C.
G c 7		**Lingula,** sp. New, larger than last, and very closely striated in lines of growth.	a. 714, Dudley, L. Ludlow?
G c 7		**Lingula,** sp. Large, flat (new?).	a. 709, Dudley, L. Ludlow.
		Obolus, Eichwald, 1829.	
G c 7		**Obolus transversus,** Salter (Davidson, Sil. Brach. p. 59, t. 5, figs. 1—6).	a. 701, a. 702, Dudley, W. Limestone and Shale; a. 703—a. 705, W. Shale, true variety, Dudley, F. C.
G c 7		**Obolus Davidsoni,** Salter (Davidson, Sil. Brach. t. 4, figs. 30—39). Muscles broad, large.	a. 705*, Dudley (limestone), F. C.
		2. *Hinged Calcareous Shells.*	
		Spirifer, Sowerby. Altered to *Spirifera* without much reason by various authors. Priority, unless a name be obviously wrong, rules in Natural History.	
F C		**Spirifer sulcatus,** Hisinger (Leth. Suecica, t. 21, fig. 6, Davidson, Sil. Br. pl. 10, figs. 4—6, Siluria, 2nd ed. pl. 21, figs. 5, 6).	a. 720, Dudley, F. C.

Case and Column of Drawers.	Reference to McCoy's Synopsis: and Figures of Genera.	Names and References; Observations, &c.	Numbers and Localities.
G d 3 F C	*Spirifera subspuria,* p. 195. 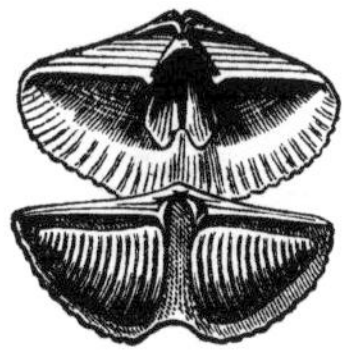	**Spirifer elevatus,** Dalm. (Davidson, Sil. Brach. p. 95, t. 10, figs. 7—11, *S. octoplicatus,* Sil. Syst. pl. 12, fig. 7, but *S. elevatus,* Salter in Mem. Geol. Surv. Vol. III. p. 278).	a. 789, interior dorsal v.; a. 790, interior ventral v.; a. 721, Dudley, F. C.
		Spirifer plicatellus, Linn. sp. (*Anomia,* Linn. See Davidson's learned discussion, Sil. Brach. p. 84, t. 9, figs. 9—12).	
F C G d 2	p. 195.	**Spirifer plicatellus,** var. **radiatus,** Sow. (Davidson, Sil. Brach. p. 87, pl. 9, figs. 1—6). *Spirifer radiatus,* Sow. Sil. Syst. pl. 12, fig. 6. *Spirifer plicatellus,* var. *radiatus,* Salter, Mem. Geol. Surv. Vol. II. pt. 1, p. 382, Siluria, 2nd ed. pl. 21, fig. 2.	b. 739, Keeper's Lodge, Golden Grove; a. 725, young; a. 726, full grown; Dudley, F. C.
G d 3		**Spirifer plicatellus,** var. **globosus,** Salter (Mem. Geol. Surv. Vol. II. pt. 1, p. 382. Davidson, l. c. pl. 9, figs. 7, 8).	a. 791 to a. 794, Dudley, F. C.
G d 3 G d 2	*Spirifera cyrtæna,* p. 193.	**Spirifer plicatellus,** var. **interlineatus,** Sow. (Davidson, l. c. p. 84, pl. 9, figs. 9—12, Siluria, 2nd ed. pl. 21, fig. 1).	a. 787, Dudley, F. C.; b. 740, Ledbury; b. 741, near Woolhope.
G d 2	p. 193.	**Spirifer crispus,** His. not Linn. (Davidson, l. c. pl. 10, figs. 13—15). Differs from *S. elevatus* chiefly in the fewer ribs and these being more deeply striate concentrically. The original figure of Linnæus evidently represents *S. sulcatus,* His. Leth. Suec. p. 73.	b. 742, Whitfield, Tortworth (Earl Ducie); b. 743, Clungunford, W. Shale.
G d 3		**Spirifer,** var. with faint ribs.	a. 795, Dudley, F. C.
F C G d 2 G d 3	*Spirifera trapezoidalis,* p. 196 (*Cyrtia*)	**Spirifer exporrectus,** Wahl. sp. (*S. trapezoidalis,* Dalman, Siluria, 2nd ed. pl. 21, fig. 3. *Cyrtia exporrecta,* Davidson, l. c. t. 9, figs. 13—24). Remarkable, even among *Spirifers,* for the extreme elevation of the beak.	a. 722, a. 723, covered by Monticulipora; a. 724, extreme variety with greatly elevated hinge; a. 788, interior parts, Dudley, F. C.; b. 744, Golden Grove, Llandeilo.

Case and Column of Drawers.	Reference to McCoy's Synopsis: and Figures of Genera.	Names and References; Observations, &c.	Numbers and Localities.
G d 1 G d 3	*Spondylobolus craniolaris,* p. 255. Pl. 1 H, figs. 4, 5.	**Athyris obovata,** Sow. sp. (Siluria, 2nd ed. pl. 22, fig. 16, Dav. Sil. Brach. p. 121, t. 12, fig. 19, t. 13, fig. 5).	a. 382, Builth Bridge (W. Shale, not Bala, J.W.S.); a. 796, interior dorsal valve; a. 797, ventral valve; a. 798, good, Dudley, F. C.
G d 3		**Meristella Circe,** Barr. (Davidson, Sil. Brach. p. 116, t. 10, figs. 33—35). Differs chiefly from *Merista,* its ally in Bohemia, by the attachment of the muscles to the ventral valve. In this they are attached directly to the shell. In the Bohemian form they are first attached to an arched shelly process, and this to the rest of the shell.	a. 660, Dudley, Wenlock Limestone, F. C.
G d 3 F C	*Athyris,* p. 196.	**Meristella tumida,** Dalman, sp. (Davidson, Sil. Brach. p. 109, t. 11, figs. 1—13, *Athyris,* Siluria, 2nd ed. pl. 22, fig. 20). Equally common in Sweden.	a. 799, good interiors; a. 661, young; a. 727 (Davidson's figure, L. Geol. Journ. fig. 12) shews spires; a. 728, a. 729, a. 730, Dudley, F. C.
G d 1 G d 3	*Hemithyris,* p. 201.	**Meristella didyma,** Dalm. (Davids. l. c. t. 12, figs. 1—10, Siluria, 2nd ed. pl. 22, fig. 15). More common in the Aymestry Limestone.	a. 800, a. 801, Dudley, F. C.; a. 781*, Ledbury; a. 781, Tortworth (Earl Ducie).
G d 3		**Meristella læviuscula,** Sow. (*Terebratula,* Sil. Syst. pl. 13, fig. 14, Siluria, 2nd ed. pl. 22, fig. 14, *M. nitida,* Davidson, Sil. Brach. t. 10, figs. 28—32). A small edition of the last.	a. 649, a. 650, Dudley, F. C.
G d 3		**Eichwaldia Capewellii,** Davidson (Sil. Brach. p. 193, t. 25, figs. 12—15, *Rhynchonella,* Salter, Siluria, 2nd ed. Foss. 57, fig. 4. *Porambonites,* Siluria, 3rd ed. p. 226). A curiously netted surface on the shell.	a. 648, Dudley, F. C.
		Retzia. A shell like *Rhynchonella,* but having the internal arm-supports wholly calcified. The shell is largely punctate.	
G c 8		**Retzia Barrandii,** Davidson (Sil. Brach. p. 128, t. 13, figs. 10—13, Siluria, 2nd ed. Foss. 57, fig. 5).	b. 745, Dudley, F. C.
F C		**Retzia Salteri,** Davidson (l. c. t. 12, figs. 21, 22, Siluria, 2nd ed. Foss. 57, fig. 7, p. 250). Easily distinguished by the centre ribs being much smaller.	a. 771, ordinary form; a. 772, fewer ribs; a. 773 approaches *R. Baylei,* Dudley, F. C.

Case and Column of Drawers.	Reference to McCoy's Synopsis: and Figures of Genera.	Names and References; Observations, &c.	Numbers and Localities.
G c 8		**Retzia Salteri,** var. **Baylei,** Davidson (l. c. t. 12, figs. 23—25). Only a longer variety than the other.	b. 747, Dudley, F. C.
G c 8		**Retzia Bouchardi,** Davidson (l. c. p. 127, t. 12, figs. 26—30, is a common species at Dudley).	
		Rhynchonella, Fischer. One recent species, the *Rhynchonella psittacea,* alone remains of this abundant fossil genus. It is common in all chalk, oolite, carboniferous and Silurian deposits; and occurred in the Upper Cambrian more rarely. Below this it is never found.	
G c 8	*Spirigerina,* p. 197.	**Rhynchonella cuneata,** Dalm. (Siluria, 2nd ed. pl. 22, fig. 8. Dav. Sil. Brach. p. 164, t. 21, figs. 7—12). A common Dudley fossil.	b. 746, Dudley, F. C.
G d 4		**Rhynchonella,** n. s. Sharp plaits, like young *Rhync. nucula.*	a. 910, a. 911, Dudley, F. C.
F C		**Rhynchonella deflexa,** Sow. (Davidson, l. c. t. 22, figs. 24—27, Siluria, 2nd ed. pl. 22, fig. 10). The shell is not really reversed; the ventral valve is so much flatter and more hollowed, and the upper valve so very convex, that it quite overhangs the other, as in several *Orthides.*	a. 769, ordinary variety, large; a. 770, with fewer ribs, Dudley, F. C.
F C	*Hemithyris,* p. 203.	**Rhynchonella Lewisii,** Davidson (l. c. t. 23, figs. 25—28, Siluria, 2nd ed. Foss. 57, fig. 2, p. 250). One of the commonest species; and yet one of the last described. The pretty fringe-like arrangement of the lines of growth, crossing the radiating ribs, renders this a very elegant shell.	a. 759, ventral valve, interior; a. 760, dorsal valve, do.; a. 761, full grown, Dudley, F. C.
F C G d 1 G d	*Hemithyris lacunosa,* p. 202.	**Rhynchonella borealis,** Schloth. (*R. lacunosa,* Dalm.), but not of Linnæus. (Salter, Mem. Geol. Surv. Vol. II. pt. 1, pl. 28, figs. 9—14.) The four plaits in the sinus, and the pointed beak, distinguish it. Davidson, Sil. Brach. p. 174, t. 21, figs. 14—20, *var.* figs. 24—27.	a. 767 (interior); a. 768, good; b. 751, Dudley, F. C.; b. 749, Ledbury; b. 750, Wenlock; Whitfield, Tortworth (Earl Ducie).

Case and Column of Drawers.	Reference to McCoy's Synopsis: and Figures of Genera.	Names and References; Observations, &c.	Numbers and Localities.
G d 1	*Hemithyris*, p. 201.	**Rhynchonella borealis,** var. **diodonta,** Dalm. (Salter, id. p. 383. Davidson, l. c. t. 21, figs. 21—23, Siluria, 2nd ed. pl. 22, fig. 5).	a. 647, Dudley, F. C.
G d 1	*Hemithyris nucula*, p. 204.	**Rhynchonella nucula,** Sow. (Siluria, 2nd ed. pl. 22, figs. 1, 2, Foss. 57, fig. 1, p. 250).	a. 381, Coed Sion, Llangadoc; b. 753, Golden Grove, Llandeilo, large (W. Shale).
F C	*Hemithyris*, p. 206. *Hemithyris crispata*, p. 200.	**Rhynchonella Stricklandii,** Sow. (Siluria, 2nd ed. pl. 22, fig. 11, Davidson, l. c. t. 21, figs. 1—6). One of the largest species, and ranges throughout the central Silurian districts, and South Wales (var. *crispata*, Dav. Sil. Brach. t. 21, fig. 28).	a. 757, Dudley, F. C.; a. 758, Malvern, Sedgley?
G d G d 1	*Hemithyris*, p. 207. *Hemithyris pentagona*, p. 205. *H. sphæroidalis*, p. 206. *H. Davidsoni*, p. 200.	**Rhynchonella Wilsoni,** Sow. (Siluria, 2nd ed. pl. 22, fig. 13, Davidson, l. c. p. 167, pl. 23, figs. 1—9). This very common and variable shell commenced life in the Woolhope period, and then was largest; dwindled away to a very poor and small size in the Wenlock limestone, and again became of magnificent proportions in Aymestry rocks.	b. 753, Dudley, F. C.; a. 380, as *H. pentagona*, Clungunford, W. Shale? a. 672, var. Falfield, Tortworth (Earl Ducie); a. 677, var., Ledbury (as *H. sphæroidalis*); a. 673, figured specimen, Dudley.
F C	*R. Davidsoni*, p. 200, in part.	**Rhynchonella Wilsoni,** var. **Davidsoni,** McCoy. This was evidently intended by the Professor to include his specific term. But it is probably only a variety. (Mr Davidson has figured it extremely well in the fifth volume of the Bull. Soc. Géol. de France, 2nd ser. pl. 3, fig. 36, as *R. sphærica*, a name previously adopted.)	a. 762 is fig. 43 of the work cited; a. 763, also figured; a. 764 is a var. with very few ribs, and a. 765 an elongated form; a. 766, interior, Dudley, F. C.
G d 1	*Hemithyris*, p. 204.	**Rhynchonella navicula,** Sow. (Siluria, 2nd ed. pl. 22, fig. 12, Davidson, Sil. Br. t. 22, figs. 20—23), is not rare in the Wenlock Shale, Dinas Bran, Llangollen, but never found at Dudley or Wenlock in the limestone.	b. 752, Wenlock Shale, Dinas Bran, Llangollen, S. Wales.
		Pentamerus, J. Sowerby, Sen. An excellent name for a shell which is literally divided into five chambers by vertical plates; so much so that the two halves of the shell come readily apart under the hammer. No Brachiopod so thoroughly shews the true	

Case and Column of Drawers.	Reference to McCoy's Synopsis: and Figures of Genera.	Names and References; Observations, &c.	Numbers and Localities.
		nature of the order, composed of bivalves which are opened along the hinge line and soldered there. Each dorsal valve therefore represents an entire bivalved shell; and the ventral valve represents the calcified foot of ordinary bivalves. (Salter, in Camb. Phil. Trans. 1869.)	
	p. 209.	**Pentamerus Knightii,** Sow. (Min. Conch. 1813, Vol. I. pl. 28, Siluria, 2nd ed. pl. 21, fig. 10, Davidson, l. c. pl. 16, 17, 19). The largest Brachiopod in the Silurian rocks. And the same species, or one very nearly allied (*P. conchidium*), ranges up to the Arctic regions, and was brought thence by the discoverers in the Arctic expedition.	Rare in Dudley limestone or shale.
F C G d 1	p. 208, as *P. globosus.*	**Pentamerus galeatus,** Dalm. (Siluria, 2nd ed. pl. 21, figs. 8, 9, Davidson, l. c. pl. 15, figs. 13—22). Certainly the most common species, and as common in Sweden and North America.	b. 754, Walsall; a. 731—a. 733, Dudley, F. C.; b. 755, Woolhope.
F C		**Pentamerus galeatus.** Strongly plaited variety. (Davidson, l. c. t. 15, fig. 19.)	a. 734, Dudley, F. C.
F C		**Pentamerus linguifer,** Sow. (Siluria, 2nd ed. pl. 22, fig. 21, Davidson, l. c. t. 17, figs. 11—14). A short broad species.	a. 735, Dudley, F. C.
		Atrypa, Dalman. Now confined to the few species (the individuals are countless) which have the calcareous spires coiled vertically, i.e. the apex pointing to the centre of the ventral valve, not horizontally as in *Spirifer*, &c. *A. reticularis*, Linn., is the type, and is world-wide, from Arctic America to Australia; and ranging from Llandovery or Upper Bala to the Upper Devonian rocks.	

Case and Column of Drawers.	Reference to McCoy's Synopsis: and Figures of Genera.	Names and References; Observations, &c.	Numbers and Localities.
F C G d 1	*Spirigerina*, p. 198.	**Atrypa reticularis,** Linn. (*Anomia reticularis*, Syst. Naturæ, 12th ed. Siluria, 2nd ed. pl. 21, figs. 12, 13, Davidson, l. c. t. 14). Strong lamellar fringes roughen the whole shell and double its bulk, but these are rarely preserved, except in shale, being of course entangled in the matrix.	a. 777, ventral valve interior; a. 778, dorsal valve interior; a. 780, Dudley, F. C.; b. 756, Wenlock; b. 757, Sedgley; Myddleton Park (Wenlock Shale), Caermarthen; b. 760, Coed Sion, Llangadoc (W. Shale); b. 761, Keeper's Lodge, Golden Grove (W. Shale).
G d 1 G d	*Spirigerina*, p. 197.	**Atrypa marginalis,** Dalm. (Siluria, 2nd ed. pl. 22, fig. 19, in pt., Davidson, l. c. t. 15, figs. 1, 2). Almost as common as the last in Silurian rocks, and ranging down to Middle Bala, but never rising above Silurian. The defined sinus in the front, and the moderately narrow striæ, distinguish it from the next form.	b. 762, Dudley, F. C.; b. 760, Wenlock.
F C	var. of *S. marginalis*, p. 197.	**Atrypa imbricata,** Sow. (Sil. Syst.) Siluria, 2nd ed. pl. 22, fig. 19, in pt. Davidson, Sil. Brach. t. 15, figs. 3—8. Coarse and few ribs, and the want of a defined medial sinus, and the rough imbrication, are characters.	a. 774, large and fine; a. 775, young variety; a. 776, very few ribs, Dudley, F. C.
G d 4		**Atrypa? Grayii,** Davidson (Sil. Brach. p. 141, t. 13, figs. 14—22, *Rhynchonella*, Siluria, 2nd ed. Foss. 57, fig. 3).	a. 651, a. 652, Walsall, Dudley, F. C.
	Hemithyris, p. 201.	**Athyris? depressa,** Sow. (Davidson, Sil. Brach. t. 13, fig. 6).	Dudley.
		Athyris? compressa, Sow. (Siluria, 2nd ed. pl. 22, fig. 22).	Dudley.
		Nucleospira, Hall. Spiral calcified arms, and a minute area in either valve, and a smooth surface, mark this obscure genus.	
G d G d 1	*Hemithyris*, p. 205.	**Nucleospira pisum,** Sow. (*Spirifer?* Sil. Syst. and Siluria, 2nd ed. pl. 21, fig. 7, Davidson, Sil. Brach. t. 10, figs. 16—20).	b. 764, Dudley, F. C.; b. 765, Wenlock.

Case and Column of Drawers.	Reference to McCoy's Synopsis: and Figures of Genera.	Names and References; Observations, &c.	Numbers and Localities.
		Orthis, Dalm. Of all Brachiopod genera the most characteristic of Lower Palæozoic rocks. Only a few forms, and those of a particular group (*O. filiaria*, *O. Michelini*) range into Carboniferous rocks. A wide hinge line, tolerably flat shells (sometimes bilobed), triangular foramen (open), and absolute want of spines on the surface, easily distinguish it from *Producta* and its allies, and the single (not double) cardinal process (i.e. *pedal muscle support*) from *Strophomena*. It is hardly ever bent as in the last-named genus, and the valves are almost equally convex in many cases; in others the dorsal, rarely the ventral, is the flat one. Only one, *O. calligramma*, of the coarse-ribbed forms, comes up from Cambrian or Cambro-Silurian rocks to mix with the striated and faintly-ribbed forms characteristic of the Silurian.	
G c 8	p. 214.	**Orthis calligramma,** Dalman. (De Vern. Geol. Russia, t. 13, fig. 7; Salter, in Mem. Geol. Surv. Vol. II. Pt. 1, p. 374, for description. Also Vol. III. pl. 22, and p. 335 to p. 337, Davidson, Sil. Brach. p. 240, pl. 35, figs. 1—17).	b. 766, Dudley, F. C.
G c 8		**Orthis,** var. **Davidsoni,** De Vern. (Bull. Soc. Geol. de France, Vol. V. 2nd ser. pl. 4, fig. 9).	b. 767, Dudley, F. C.
F C	p. 226.	**Orthis rustica,** Sow. (Sil. 2nd ed. pl. 20, fig. 10. Davidson, Lond. Geol. Journ. p. 64, pl. 13, figs. 1—4).	a. 741, dorsal valve; a. 742, ventral valve, Dudley, F. C.
G d 2	do.	**Orthis rustica,** var. **rigida,** Davidson (Lond. Geol. Journ. p. 63, pl. 13, figs. 16, 17), evidently an irregularly grown variety of the last.	b. 768, Dudley.
		Orthis, var. **Walsalliensis,** Davidson (Bull. Soc. Géol. de France, 2nd ser. Vol. V. pl. 4, fig. 7). I think only a many-ribbed variety of *O. rustica*.	Dudley.
F C	p. 213.	**Orthis biloba,** Linn. (*Ter. sinuata*, J. Sowerby. *Delthyris cardiospermiformis*, von Buch). One of the prettiest fossils of the limestone, and most instructive as shewing (with *Terebratula diphya* and others) the double-valved nature of the dorsal valve in Brachiopods. (Siluria, 2nd ed. pl. 20, fig. 14. Davidson, Brit. Fos. Brach. Introd. pl. 8, fig. 141.)	a. 755, dorsal valve; a. 756, ventral, Dudley, F. C.

Case and Column of Drawers.	Reference to McCoy's Synopsis: and Figures of Genera.	Names and References; Observations, &c.	Numbers and Localities.
F C G d 2	p. 216.	**Orthis elegantula,** Dalm. (Siluria, 2nd ed. pl. 20, fig. 12. Davidson, Sil. Brach. t. 27, figs. 1—9). Not so common in the Dudley limestone as in the Bala rocks, but still very frequent. I think *O. parva* a mere variety, and feel inclined to keep *O. orbicularis* distinct.	a. 753, interior ventral valve; a. 752, interior dorsal valve; a. 754, mere trigonal variety, Dudley, F. C.; b. 769, Erw Gilfach, Builth; a. 785 (as *O. turgida*), Coed Sion, Llangadoc, W. Shale.
	O. parva, p. 221.	**Orthis,** var. small and triangular.	
G d		**Orthis orbicularis,** Sow. (Siluria, 2nd ed. pl. 20, fig. 9). A much rounder and less triangular shell than *O. elegantula,* but still very closely allied. It is far more common in Ludlow rocks (as at Sedgley), but here and there occurs in true Wenlock; as at Benthall Edge.	a. 892, a. 893, a. 894, Falfield, Tortworth (Earl Ducie).
G d G d 2	p. 220.	**Orthis hybrida,** Sow. (Siluria, 2nd ed. pl. 20, fig. 13. Davidson, Sil. Brach. p. 214).	b. 770, Dudley, F. C.; b. 771, Whitfield, Tortworth (Earl Ducie).
G d 4 G c 8		**Orthis Lewisii,** Davidson (Bull. Soc. Géol. Fr. 2nd. ser. Vol. v. t. 3, fig. 19, Sil. Brach. t. 26, fig. 4).	a. 659, Dudley, F. C.
	p. 223.	[**Orthis porcata,** McCoy (see p. 89), is found, though rarely, in the Dudley limestone.]	
G c 8 G d		**Orthis Bouchardi,** Davidson (London Geol. Journ. pl. 13, figs. 5—8; Bull. Soc. Géol. Fr. Vol. v. pl. 3, fig. 19).	b. 748, b. 772, Dudley, F. C.
G d	*Spirifera,* p. 192.	**Orthis biforata,** Schloth (De Vern. Geol. Russ. t. 3, Davids., l.c. Introduction, pl. 8, fig. 146). Rare in Wenlock rocks: common in Cambrian rocks:—this species lingers on, to shew among many others how much closer the relations are between the Upper Cambrian and the true Silurian strata, than between the former and the lower Cambrian.	b. 773, Dudley, F. C.
G c 8		**Orthis æquivalvis,** Davidson (Sil. Brach. t. 30, figs. 9, 10; Siluria, 2nd ed. p. 251).	b. 774, Walsall, W. Shale.

Case and Column of Drawers.	Reference to McCoy's Synopsis: and Figures of Genera.	Names and References; Observations, &c.	Numbers and Localities.
	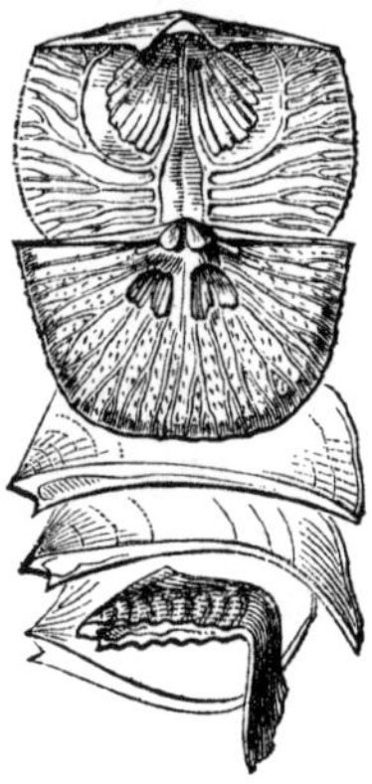	**Strophomena,** Rafinesque. Easily distinguished from *Orthis* by the broad expanded form —one valve being often bent on the other, and so flat as to leave very little room for the animal. The general characters are much the same as in *Orthis*. But the cardinal process is invariably double, not single. *Leptagonia,* McCoy, is only a *Strophomena.* And the Professor has placed all this genus, natural as it is, under the general term *Leptæna,* Dalman, which is best restricted, as Davidson and myself have done, to the involute species with long, instead of square dorsal muscles.	
F C **G c**	*Leptæna,* p. 243.	**Strophomena euglypha,** Dalman (Siluria, 2nd ed. pl. 20, fig. 19). Davidson, Lond. Geol. Journ. Vol. I. pl. 12, figs. 1—4?). Most common. The dorsal valve bends down over the ventral, which is bent backwards.	b. 781, a. 743, interior of dorsal valve; a. 744, a. 745, outside of ventral valve; a. 746, outside of dorsal valve, Dudley, F. C.
G c 8	*Leptæna,* p. 244.	**Strophomena funiculata,** McCoy (*Orthis,* Sil. Foss. Ireland, pl. 3, fig. 11. Davidson, Lond. Geol. Journ. t. 12, fig. 6, Siluria, 2nd ed. p. 251). Bent in the reverse way to *S. euglypha,* and a miniature copy of that species.	b. 775, Dudley, F. C.
G d	*Leptæna,* p. 236.	**Strophomena Ouralensis,** De Vern. (Geol. Russ. t. 14, fig. 1, *S. imbrex,* Davidson, and Salter, Siluria, Foss. 58, fig. 6, not of Pander). Gently curved, the ventral valve the convex one. This differs from Pander's *O. imbrex,* which is narrower, and abruptly bent down. The interior is beautifully shewn in Davidson's figures.	b. 776, Dudley, F. C.
G d 2 **G c 9**	*Leptæna,* p. 243.	**Strophomena filosa,** Sow. sp. (*Orthis,* Sil. Syst. t. 13, fig. 12, Siluria, 2nd ed. pl. 20, fig. 21, Davidson, Lond. Geol. Journ. Vol. I. p. 62). A common flat species, beautifully striate.	b. 778, Dudley, F. C.; b. 779, Keeper's Lodge, Golden Grove, Llandeilo.
G c **G d 2**	*Leptæna,* p. 245.	**Strophomena Pecten,** Linn. sp. (*Anomia,* Syst. Naturæ, *Orthis Pecten,* Siluria, 2nd ed. Foss. 58, fig. 3. Davidson, Introd. Foss. Brach. t. 8, fig. 163, &c.). Very common in Silurian and Upper Cambrian rocks.	b. 780, Dudley, F. C.; b. 783, Myddleton Park, Caermarthen.

Case and Column of Drawers.	Reference to McCoy's Synopsis: and Figures of Genera.	Names and References; Observations, &c.	Numbers and Localities.
F C G c G d 2	*Leptagonia,* p. 248. (as *L. deltoidea β,* p. 234.)	**Strophomena depressa,** Dalm. (*Anom. rhomboidalis,* Wahl. *Leptæna,* Sil. Syst. pl. 12, figs. 12—16. Mem. Geol. Surv. Vol. II. Part 1, p. 283. *Strophomena,* Siluria, 2nd ed. pl. 20, fig. 20).	a. 784 (as *L. deltoidea β undata*), Coed Sion, Llangadoc; a. 736, a. 737, interior dorsal valve; a. 738—a. 740, interior ventral valve, Dudley, F. C.
G d 2	*Leptæna,* p. 233.	**Strophomena corrugata,** Portl. sp. (Geol. Rep. Lond. p. 450, t. 32, figs. 17, 18, *non* Conrad, Jour. Ac. Nat. Sc. Phil. Vol. VIII. p. 256, t. 14, f. 8. *S. corrugatella,* Dav. Sil. Brach. p. 301, t. 41, fig. 8).	a. 783, Golden Grove, Llandeilo, W. Shale; Myddleton Park?
G d 2 G c 9		**Strophomena quadrata,** Lindström. Much smaller in all its parts; and with the abrupt front turned down only a short way. It is oblong, compared with its larger congener, and more delicately striated.	b. 777, a. 715, Dudley, F. C.
G c 8 G d 2 G c 9	*Leptæna,* p. 241.	**Strophomena antiquata,** Sow. (Siluria, 2nd ed. pl. 20, fig. 18, Foss. 58, fig. 8, Davidson, Sil. Brach. p. 297, t. 44, fig. 2). A rough, but elegant shell, coarser in its imbricated striation than any other *Strophomena.*	a. 882, Walsall; b. 783, b. 784, Dudley, F. C.
G b	*Leptæna,* 1 H, figs. 33—35, p. 246.	**Strophomena simulans,** McCoy. A species I fear founded on several specimens of several species. I cannot define it.	a. 465, Myddleton Park, Caermarthen, W. Shale.
G c 9		**Strophomena,** sp.	Dudley, F. C.
		**Leptæna,** Dalman, proper. Shells involute. The ventral enveloping the dorsal one. The cardinal processes connate with the widely set hinge-teeth (in *Strophomena* they are distinct), and the muscular scars very long —not squarish. The habit of the three genera *Orthis, Strophomena, Leptæna,* is so distinct, that little difficulty can be found by the student in separating them.	
G d 2 F C	p. 240. as *L. sericea,* var. *rhombica,* p. 239. as *L. quinquecostata,* p. 236.	**Leptæna transversalis,** Dalman (Siluria, 2nd ed. pl. 20, fig. 17, Davidson, l. c. t. 48, fig. 1). One of the shells that range through a good many formations (Bala rocks to	a. 782, as *L. sericea,* var. *rhombica,* Llyn Alwen; a. 747, outside dorsal valve; a. 748, interior ditto; a.

Case and Column of Drawers.	Reference to McCoy's Synopsis: and Figures of Genera.	Names and References; Observations, &c.	Numbers and Localities.
F C		Ludlow rocks) and yet preserve their characters unimpaired. *Leptæna Duvalii* is one of its extreme Dudley forms. It is equally common in sandy, muddy and limestone deposits; but appears to prefer the calcareous mud.	749, outside ventral valve; a. 750, interior (cast) ditto; a. 751, young; Dudley, F. C.
		Chonetes, Fischer. All the *Productidæ*, to which this shell belongs, have spines in some part of the surface. *Chonetes* has them confined to the hinge-line of the ventral valve. Carboniferous species are large and finely striated. Devonian ones coarsely ribbed, often. The Silurian forms are small, and often striate roughly or are smooth, sometimes they are minute shells.	
G d 2	p. 249.	**Chonetes lata,** Von Buch sp. (Siluria, 2nd ed. pl. 20, fig. 8, *C. striatella*, Dalm. sp. Davidson, Sil. Brach. t. 49, figs. 23—26). Of all the common Ludlow shells this is the commonest; but it is rarely found in Wenlock limestone, and occasionally reaches May Hill sandstone. It has coarse striæ and spines along the hinge, and thus is easily distinguished from the species of *Leptæna* proper.	b. 789, Myddleton Park, Caermarthen.
G d 2	*Leptæna*, p. 235.	**Chonetes lævigata,** Sow. (*Leptæna lævigata*, Sow. Sil. Syst. Siluria, 2nd ed. pl. 20, fig. 15). Smooth, with extended ears, and extremely common in muddy Wenlock Shales. McCoy has occasionally mistaken this for other species (*Orthis parva* for instance).	b. 790, Clungunford, Shropshire; b. 791, near Pool; b. 792, Keeper's Lodge, Golden Grove.
G d 2	*Leptæna*, p. 235.	**Chonetes minima,** Sow. (*Lept. minima*, Siluria, 2nd ed. pl. 20, fig. 16).	b. 794, W. Shale, Llanfair, Welchpool; b. 795, Bed of Dee, Llantysilio.
G c 8		**Chonetes minima,** Sow. *var.* Grayii, Davidson (Sil. Brach. p. 334, t. 49, figs. 15—19), *Leptæna Grayii*, Davidson (Bull. Soc. Géol. Fr. 2nd ser. Vol. v.).	b. 788, Dudley, F. C.

WENLOCK CONTINUED.

LAMELLIBRANCHIATA (*Conchifera* of Authors).

Bivalve shells proper: differ from Brachiopods not only in the want of the spirally coiled and ciliated arms—the tentacles, four in number, being simple and flaccid—but in wanting the ventral shelly cover to the foot, which in the Brachiopod becomes the ventral valve. The foot on the contrary, though often giving birth to a byssus, is a free organ, useful for locomotion, and capable of great and varied evolutions. The valves in the infant state are open (as Rathke has shewn in Anodon), and the heart and intestine double (the heart is permanently double in the Brachiopod), and these coalesce as the animal grows and closes the valves. The muscles instead of all being directed to the foot and its appendages as in the *Terebratula,* are chiefly used for transverse action to close the valves, a portion only being directed to the important foot. The higher genera, not requiring to be anchored (*Cardium, Venus,* &c.), do not spin a *byssus.* This, which arises from a large gland in the back of the foot, is represented by the anchor or plug of the *Terebratula;* and, as I have lately shewn (Camb. Philos. Trans.), the operculum of the Gasteropod is an analogous organ. The *Bivalves* then stand half way between the *snails* and *whelks,* with free motion, and the permanently fixed and helpless *Terebratula.*

It is important to observe, and Professor Phillips was the first to point it out (Mem. Geol. Surv. Vol. II. Pt. 1, p. 264), that the lower genera only of Lamellibranchs—*Arca* and its allies, *Avicula, Pterinea, Modiola, Mytilus,* to which may be added *Nucula,* the freest of this group—are present in the earliest rocks in which fossils are found. To these may be added *Conocardium,* which, like *Teredo* and *Pholadidea* in our seas, have the valves soldered.

Case and Column of Drawers.	Reference to McCoy's Synopsis: and Figures of Genera.	Names and References; Observations, &c.	Numbers and Localities.
F C	Pl. 1 I, figs. 11—15, p. 258.	**Avicula Danbyi,** McCoy (Siluria, 2nd ed. Foss. 59, figs. 2, 3). A common shell in the Ludlow Sandstones of Kendal.	a. 803, Dudley, F. C.[1]
F C		**Avicula mira,** Barr. MS. (Siluria, 2nd ed. p. 253).	a. 820, a. 821, a. 822, left valve; a. 823, right valve; a. 824, both valves united, rare, Dudley, F. C.
F C	p. 263.	**Pterinea Sowerbyi,** McCoy (Siluria, 2nd ed. pl. 23, fig. 15. *Avicula reticulata,* Sil. Syst. t. 6, fig. 3). A long direct shell, not transverse as in many species.	a. 818, right valve, Dudley, F. C.
G d 7	Pl. 1 I, fig. 5, p. 259.	**Pterinea asperula,** McCoy (Siluria, 2nd ed. p. 253, Foss. 59, fig. 4).	b. 720, Builth, W. Shale.
G d 7 **G e 5**		**Pterinea lineatula,** D'Orb. (Siluria, 2nd ed. pl. 23, fig. 16 *Avicula lineata,* Sow., is not the *Avicula lineata* of Goldfuss).	b. 727, Myddleton Park, Caermarthenshire (Wenlock Shale); b. 728, Dudley, F. C.

[1] F. C. Fletcher Collection.

Case and Column of Drawers.	Reference to McCoy's Synopsis: and Figures of Genera.	Names and References; Observations, &c.	Numbers and Localities.
F C	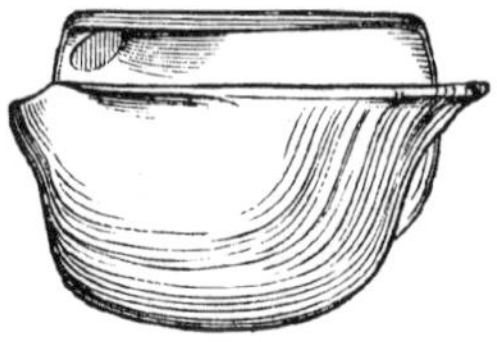	**Pterinea,** new like *lineatula.*	a. 804, Dudley, F. C.; ? Wenlock Shale or Lower Ludlow.
F C		**Pterinea? planulata,** Conrad (Siluria, 2nd ed. Foss. 59, fig. 6). A common shell, but the valves are so equal that it can hardly be a *Pterinea.*	a. 839, anterior and posterior ear good, Dudley, F.C.
F C	*Pterinea.* 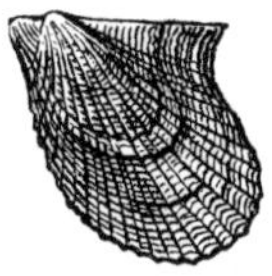	**Pterinea** (sp. 2) **exasperata,** Salter, MSS. Long known, but not yet described. A reticulate species.	a. 813, a. 814, a. 816, good left valve; a. 815, interior of same valve, Dudley, F. C.
		Pterinea (sp. 2) var. coarser ridges.	a. 817, Dudley.
F C	p. 262.	**Pterinea retroflexa,** His. (Siluria, 2nd ed. pl. 23, fig. 17). Widely distributed over N. Europe. The ridges vary much and the form too.	a. 805, a. 806, left valve; a. 807, right valve, Dudley, F. C.
F C		**Pterinea,** var. coarse imbrications.	a. 819, Dudley, F. C.
G d 7	Pl. 1 I, fig. 4, p. 263.	**Pterinea tenuistriata,** McCoy (Siluria, 2nd ed. Foss. 59, fig. 5).	b. 721, Myddleton Park, Caermarthen; b. 721*, Erw Gilfach, Builth.
G d 7	Pl. 1 I, fig. 7, p. 260.	**Pterinea demissa,** Conrad, var. of *P. retroflexa.* This seems to differ only in the squarer shape, and the lines of growth.	b. 722, Myddleton Park, Caermarthen.
	p. 266.	**Modiolopsis antiqua,** Sow. sp. (Siluria, 2nd ed. pl. 23, fig. 14). A very narrow anterior end, the lobe small.	Dudley.
G b	*M. Nilssoni,* p. 267.	**Modiolopsis gradatus,** Salter (*Mytilus,* Salter, Mem. Geol. Surv. Vol. II. Pt. 1, pl. 20, fig. 4). Sharp sudden imbrications of growth, the rest smooth.	b. 723, Myddleton Park, Caermarthen.
F G d 7		**Modiolopsis mytilimeris,** Conrad (Siluria, 2nd ed. Foss. 60, fig. 6). A very broad species—almost the shape of *Ambonychia.*	a. 811, right valve, a. 812, left valve, Dudley, F. C.; b. 724, Wenlock.

Case and Column of Drawers.	Reference to McCoy's Synopsis: and Figures of Genera.	Names and References; Observations, &c.	Numbers and Localities.
F C	p. 264. 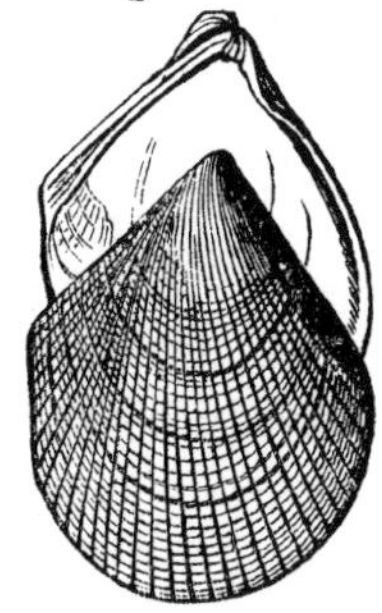	**Ambonychia striata,** Sow. (*Cardiola,* Siluria, 2nd ed. pl. 23, fig. 13).	a. 840, Dudley, F. C.
G d 7		**Lunulacardium,** Münster. A convenient genus, in which are placed a number of thin triangular bivalves with a truncated posterior side, divided by a strong keel from all the rest of the shell. Radiating striæ cover the shell.	b. 725, Builth Bridge, Wenlock Shale.
	(*Anodontopsis securiformis*), p. 272. Pl. 1 L, fig. 9.	**Pseudaxinus,** Salter. A generic term, intended to include those very thin and edentulous shells which have the shape of *Axinus.* A single species of it only is known in Wenlock and Ludlow rocks; but it occurs throughout Lower Palæozoic, and my type was described from *Arenig* rocks. No specimen in the collection.	
F C		**Pleurorhynchus æquicostatus,** Phill. (Siluria, 2nd ed. Foss. 59, fig. 1, Phill. Mem. Geol. Surv. Vol. II. Pt. 1, pl. 16). A small species—the American rocks hold much larger species.	a. 829, Dudley, F. C.
	Leptodomus, p. 278. 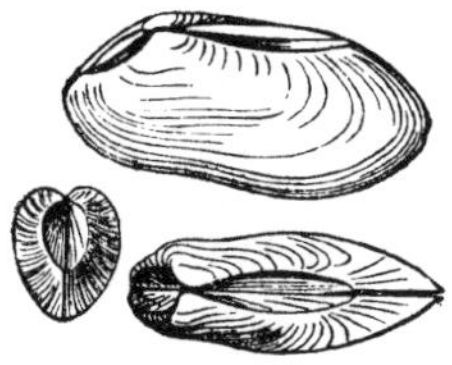	**Orthonotus amygdalinus,** Sow. sp. (Siluria, 2nd ed. pl. 23, fig. 6).	Dudley.
F C	*Leptodomus,* p. 279.	**Orthonotus impressus,** Sow. sp. (Siluria, 2nd ed. pl. 23, fig. 3).	a. 828, Dudley, F. C.
F C		**Goniophora grandis,** Salter. Strongly costated, and marked with lines of growth which decussate the ribs. Twice the size of the common Ludlow species.	a. 827, Dudley, F. C.
F C	p. 280.	**Grammysia cingulata,** Hisinger (Siluria, 2nd ed. Foss. 60, fig. 1). The genus is Silurian and Lower Devonian, allied to *Orthonotus.*	a. 836, left valve; a. 837, right valve; a. 838, young, Dudley, F. C.

Case and Column of Drawers.	Reference to McCoy's Synopsis: and Figures of Genera.	Names and References; Observations, &c.	Numbers and Localities.
G d 7		**Cardiola interrupta,** Brod. (Siluria, 2nd ed. pl. 23, fig. 12). A shell universal in Wenlock rocks throughout Europe.	b. 797, Erw Gilfach, Builth; b. 798, Maen Goran, Builth.
F C	Genus, p. 273.	**Clidophorus,** sp. McCoy (oval shells). The genus is one of the *Arcaceæ,* the teeth of which lie parallel to the hinge border. The shells are thin.	a. 826, Dudley, F. C.
G d 5 G d 7	Pl. 1 K, fig. 9, p. 273.	**Clidophorus planulatus,** Conrad, sp. (Hall, Pal. New York, t. 82, fig. 9).	a. 845, Shale, Dudley, F.C.; a. 915, Keeper's Lodge, Golden Grove.
F C		[**Actinodonta,** (Phill.) sp. Related closely to *Clidophorus,* and possibly included in it. Oval shells, not thick as most *Arcaceæ,* and with teeth on both sides lying parallel to the hinge-plate; and a few central radiating ones. A species very common in the May Hill sandstones of Marloes Bay.]	a. 830, Dudley, F. C.
G d 5	p. 283.	**Cuculella,** oval sp. like *C. antiqua* of the Ludlow rocks. The shells of this genus are endless in minute variation in all Silurian rocks, rare in Cambrian.	b. 799, Dudley, F. C.
G d 7	(*Nucula,* of many authors.)	**Ctenodonta,** Salter. Intended to replace *Nucula* for those abundant species in old rocks, all of which have *external* ligaments.	b. 800, Dinas Bran, Llangollen.
F C	*Nucula,* p. 285.	**Ctenodonta anglica,** Sow. (Siluria, 2nd ed. pl. 23, fig. 10). The hinge-lines are very much bent interiorly in this broad triangular species.	a. 835, Dudley, F. C.
G d 5		**Ctenodonta,** sp. 1.	a. 832, Dudley, F. C.
G d 5		**Ctenodonta,** sp. 2.	a. 833, Dudley, F. C.
G d 5		**Ctenodonta,** sp. 3.	a. 834, Dudley, F. C.
G d 5		**Ctenodonta,** sp. 4. A wide triangular form, like many Bala species in foreign lands, but not usual for Britain.	a. 834*, Dudley, F. C.

Case and Column of Drawers.	Reference to McCoy's Synopsis: and Figures of Genera.	Names and References; Observations, &c.	Numbers and Localities.
	Sea-Butterflies.	**PTEROPODA.**	
		Theca, Morris. A straight sheath-like shell with a flat ventral surface, and a convex dorsal one. It has an operculum. See p. 66.	
Gd 7	p. 287.	**Theca Forbesii,** Sharpe (Quart. Geol. Journ. Vol. II. pl. 13, fig. 1). A beautiful little fossil. The back convex, the front flat. This is opposite to the character of the Bala species *T. reversa.*	b. 726, Dinas Bran, Llangollen.
		Conularia, Sow. The largest and most elegant of ancient Pteropods; which order was of greater importance in Cambrian and Silurian times than it has ever been since. Some specimens were at least a foot long!	
F C Gd 6	p. 287.	**Conularia Sowerbyi,** Defr. (*C. cancellata,* Sandberger, Siluria, 2nd ed. pl. 25, fig. 10). The graceful bend of the crenated ridges across the face of each side distinguishes this beautiful and common shell. The section is rhomboidal. The carboniferous species *C. quadrisulcata* much resembles it.	a. 871, a. 872, huge specimen, Dudley, F. C.
Gd 6		**Conularia clavus,** Salter, n. s. New square sp. It is rare to find species with square section; usually the opposite angles are always equal, but the section is often rhomboidal; and generally two of the sides are smaller than the rest.	a. 878, Dudley, F. C.
		GASTEROPODA.	
	Spiral Univalves.	The great difference between the univalve and bivalve shells lies in the fact, that in the inner side of the spiral curve, i. e. on the right side of the animal, the gill is suppressed. In *Patella,* however, both gills are present, and Gasteropods are higher, for they have a head, often eyes—and some species, e.g. all the snails, breathe air.	

Case and Column of Drawers.	Reference to McCoy's Synopsis: and Figures of Genera.	Names and References; Observations, &c.	Numbers and Localities.
	Capulus, p. 290.	**Acroculia,** Phillips. More spiral than *Capulus*, and often spinose or tubercular.	
G c F C	*Capulus?* p. 290.	**Acroculia haliotis,** Sow. (Siluria, 2nd ed. pl. 24, fig. 9). A most characteristic Wenlock shell—the food of Crinoids, especially of *Marsupiocrinus.*	b. 804, Ledbury; a. 854, regular natica-like growth, a. 851, a. 855, mouth, a. 852, narrow var., i.e. slow-growing spire, a. 853, quicker growth, a. 856, Dudley, F. C.
G d 6		**Acroculia,** sp. 2.	b. 803, Dudley, F. C.
G d 6 G d 7		**Acroculia,** sp. 3. Very much angulated whorls, a small species.	a. 625, Tortworth (Earl Ducie); a. 626, Dudley, F.C.
G d 6		**Acroculia,** sp. 4.	b. 801, Dudley, F. C.
G d 6		**Acroculia prototypa,** Phil. (Siluria, 2nd ed. pl. 24, fig. 8). Very much like a *Nerita* in look.	b. 802, Dudley, F. C.
		Pleurotomaria, Sow. A genus allied to *Scissurella,* and has the mouth deeply notched. It is chiefly palæozoic, i.e. the species with a convex base.	
F C		**Pleurotomaria undata?** Sow. (Siluria, 2nd ed. pl. 24, fig. 6).	a. 846, Dudley, F. C.
F C		**Pleurotomaria,** minute sp.	a. 847, Dudley, F. C.
F C		**Pleurotomaria Fletcheri,** Salter, n. sp.	a. 851, Dudley, F. C.
F C	p. 292?	**Pleurotomaria undata,** Sow.? (*Pl. undata,* Sow. Siluria, 2nd ed. pl. 24, fig. 6.)	a. 856, Dudley, F. C. (?Lower Ludlow); it is common there.
F C		**Pleurotomaria balteata,** Phill. (Mem. Geol. Surv. Vol. II. pt. 1, p. 358, pl. 15, figs. 1, 2). The spire of this large shell is much depressed: the band rough and prominent.	a. 857, good figured specimen, Dudley, F. C.

Case and Column of Drawers.	Reference to McCoy's Synopsis: and Figures of Genera.	Names and References; Observations, &c.	Numbers and Localities.
F C		**Pleurotomaria,** sp. (depressed spire, but rounded whorls; an unusual form for the genus, which has the whorls mostly angulate).	a. 858, Dudley, F. C.
G d 5		**Pleurotomaria,** sp. 1.	b. 805, Dudley, F. C.
G d 5		**Pleurotomaria,** sp.	b. 806, Dudley, F. C.
G d 6		**Pleurotomaria,** sp. Like *P. striatissima.*	a. 876, Dudley, F. C.
G d 6		**Pleurotomaria** uniformis, Salter, n. sp., large, quite without ridges except band.	a. 879, Dudley, F. C.
		Murchisonia, De Vern. and D'Archiac.	
G d 5	p. 293.	**Murchisonia Lloydii,** Sow.? (Siluria, 2nd ed. pl. 24, fig. 5.)	b. 810, Dudley, F. C.
F C		**Murchisonia cyclonema,** Salter.	a. 848, old; a. 849, young; a. 850, shews mouth; Dudley, F. C.
		Murchisonia, Section *Hormotoma* (Salter), bead-like forms.	
G d 6		**Murchisonia,** sp. 1. Quickly tapering, two inches long.	b. 807, Dudley, F. C.
G d 6		**Murchisonia,** sp. 2. Like *H. cingulata.*	b. 808, Dudley, F. C.
G d 6		**Murchisonia,** sp. 3.	b. 809, Dudley, F. C.
		Cyclonema, Hall.	
F C		**Cyclonema cirrhosa,** Sow. (Siluria, 2nd ed. pl. 24, fig. 10, *Eunema,* 3rd ed. pl. 24, fig. 10).	a. 868, Dudley, F. C.
G d 6		**Cyclonema,** short, broad, sp. 1.	a. 873, Dudley, F. C.
G d 6		**Cyclonema,** sp.	b. 811, Dudley, F. C.

Case and Column of Drawers.	Reference to McCoy's Synopsis: and Figures of Genera.	Names and References; Observations, &c.	Numbers and Localities.
G d 6	*Litorina*, p. 305.	**Cyclonema octavia,** D'Orb. (Siluria, 2nd ed. pl. 24, fig. 4. *Turbo carinatus,* Sil. Syst. t. 5, fig. 28).	a. 874, Dudley, F. C.
		Eunema, Salter. Of frequent occurrence in Silurian and Cambrian rocks. Species are sure to be found. The shells are thin, and probably had no operculum. They resemble *Cyclonema* (Geol. Surv. Can. Decad. 1, p. 29).	
		Trochonema, Salter (Canadian Decades, No. 1). Shells like *Eunema* and *Cyclonema,* but with a wide open umbilicus, *Cyclonema* has none.	
G d 6		**Trochonema bijugosa,** Salter (n. sp.). A species much resembling the *T.* (*Turbo*) *trochleatus* of McCoy and Hall.	a. 875, Dudley, F. C.
		Macrocheilus? (*Polyphemopsis,* Portlock). Ovate shells, with closely pressed narrow whorls, looking more like a *Bulimus* than anything else. They are found in Cambrian as well as Silurian rocks.	
F C		**Machrocheilus pupa,** n. sp., Salter.	a. 869, Dudley, F. C.
		Euomphalus, Sow.	
F C G c	p. 297.	**Euomphalus centrifugus,** Wahl. sp. (Hisinger, Leth. Suec. t. 12, fig. 1).	b. 821, Wenlock; a. 841, Dudley, F. C.
F C G c	p. 298.	**Euomphalus discors,** Sow. (Siluria, 2nd ed. pl. 24, fig. 12).	a. 864, large upper side; a. 865, under side, Dudley, F.C.; b. 812, Wenlock.
G d 5		**Euomphalus,** varieties of ditto.	Dudley, F. C.
F C		**Euomphalus pacificatus,** Salter.	a. 861, Dudley, F.C.
F C		**Euomphalus Mariæ,** Salter, n. sp. Related to *E. discors,* but with most regular ridges of growth. A beautiful shell, dedicated to a most worthy lady—the patient preparer of this Collection. Two or three hours daily were given by Mrs Fletcher to this task, and the result is an unique cabinet of species new to science.	a. 859, upper side; a. 860, under side, Dudley, F. C.

Case and Column of Drawers.	Reference to McCoy's Synopsis: and Figures of Genera.	Names and References; Observations, &c.	Numbers and Localities.
G d 7 F C G c	p. 298.	**Euomphalus funatus,** Sow. (Siluria, 2nd ed. pl. 25, fig. 3).	b. 814, a. 802, Dudley, F.C.; b. 813, Whitfield, Tortworth (Earl Ducie).
F C G d 7	p. 299.	**Euomphalus sculptus,** Sow. (Siluria, 2nd ed. pl. 25, fig. 2).	a. 863, good, Dudley, F.C.; b. 815, Whitfield, Tortworth (Earl Ducie).
F C G c G d 7	p. 298.	**Euomphalus rugosus,** Sow. (Siluria, 2nd ed. pl. 24, fig. 13).	a. 866, a. 867, Dudley, F.C.; b. 816, Wenlock; b. 817, Woolhope.
F C		**Euomphalus alatus,** Sow. (Siluria, 2nd ed. pl. 25, fig. 4).	a. 842, young (*subundulatum,* Salter); a. 843, upper side; a. 844, under side, Dudley.
F C		**Euomphalus Calyptræa,** n. sp. Regular oblique coarse ridges, and angle to lower edge of whorl.	a. 862, Dudley, F. C.
G d 7	Pl. 1 K, figs. 37, 38, p. 299.	**Euomphalus triporcatus,** McCoy (probably a *Trochonema*).	a. 917, Golden Grove, Llandeilo, figured specimen.
G d 5		**Euomphalus,** n. sp.	b. 818, Dudley, F. C.
G d 6		**Euomphalus carinatus?** Sow. (Siluria, 2nd ed. pl. 24, fig. 11).	b. 836, Dudley, F. C.
	p. 303.	**Holopella,** McCoy.	
	p. 302.	**Loxonema,** Phill.	
G d 6		**Loxonema,** sp. 1. Long, like *Murchisonia,* possibly a *Hormotoma.*	b. 819, Dudley, F. C.
G d 6		**Loxonema,** *minute* sp.	b. 820, Dudley, F. C.
		HETEROPODA.	
G d 6	p. 309.	**Bellerophon dilatatus,** Sow. (section *Bucania,* Siluria, 2nd ed. pl. 25, figs. 5, 6).	a. 877, Dudley, F. C.
		Bellerophon expansus, Sow. (Siluria, 2nd ed. pl. 25, fig. 8).	
F C		**Bellerophon trilobatus,** Sow. (Siluria, 3rd ed. Foss. 15, fig. 9).	a. 870, largest specimen known, Dudley, F. C.

Case and Column of Drawers.	Reference to McCoy's Synopsis: and Figures of Genera.	Names and References; Observations, &c.	Numbers and Localities.
G d 7 G d 6	p. 311.	**Bellerophon Wenlockensis,** Sow. (Siluria, 2nd ed. pl. 25, fig. 7).	a. 627, Woolhope; b. 823, Dudley, F. C.
	Nautili, or Cuttle-fish with shells.	**CEPHALOPODA.**	
G d 8	p. 313.	**Orthoceras angulatum,** Wahl. (Hisinger, Leth. Suec. t. 10, fig. 1, Siluria, 2nd ed. pl. 28, fig. 4).	b. 824, Builth Bridge (? the true species).
	p. 314.	**Orthoceras dimidiatum,** Sow. (Siluria, 2nd ed. pl. 28, fig. 5).	
G c F C G b G d 8	(*Cycloceras*), p. 319.	**Orthoceras annulatum,** Sow. (Siluria, 2nd ed. pl. 26, fig. 1).	b. 826, a. 620, a. 621, Dudley, F. C.; b. 828, Dermydd Fawr, Craig Bronbanog, Denbighshire; b. 827, Ledbury.
F C G d 10		**Orthoceras,** var. **fimbriatum,** Sow. (Siluria, 2nd ed. pl. 26, fig. 2).	a. 618, a. 646, very coarse ridges, Dudley, F. C.
G d 10	p. 313.	**Orthoceras Brightii,** Sow. (*Ormoceras,* Siluria, 2nd ed. pl. 27, figs. 5, 6), is interior of *O. fimbriatum.*	a. 640, Ledbury.
G d 10	*O.* (*Actinoceras*), p. 315.	**Orthoceras Mocktreense,** Sow. (Sil. 2nd ed. pl. 29, fig. 2). The siphon of this is bead-like.	a. 639 (beaded siphon), Ledbury.
	p. 315.	**Orthoceras Ludense,** Sow. (Siluria, 2nd ed. pl. 28, figs. 1, 2, Foss. 61, fig. 2).	
F C		**Orthoceras canaliculatum,** Sow. (Siluria, 2nd ed. pl. 28, fig. 3).	a. 615, Dudley.
G b	p. 317.	**Orthoceras subundulatum,** Portl. (Geol. Rep. Lond. t. 28, fig. 2, Siluria, 2nd ed. Foss. 61, fig. 3).	b. 832 (Wenlock Shale), Wenlock.
G d 8	p. 317.	**Orthoceras tenuicinctum,** Portl. (Geol. Rep. p. 371, t. 27, fig. 5; Siluria, 2nd ed. p. 219, Foss. 40, fig. 3, *Creseis Sedgwicki,* Forbes).	b. 825, Dinas Bran, Llangollen.
		Orthoceras? perelegans, Salter (Siluria, 2nd ed. pl. 29, fig. 6), very common.	
G d 8	*Cycloceras,* p. 319.	**Orthoceras Ibex,** Sow. (Siluria, 2nd ed. pl. 29, figs. 3, 4).	b. 831, Dinas Bran, Llangollen.

Case and Column of Drawers.	Reference to McCoy's Synopsis: and Figures of Genera.	Names and References; Observations, &c.	Numbers and Localities.
G d 8	*O. filosum*, p. 314.	**Orthoceras dulce,** Barrande (Sil. Syst. Boh. Vol. II. pl. 294, 295; pl. 357, figs. 8, 9).	a. 628, Builth Bridge, large specimen.
G d 8	p. 315.	**Orthoceras laqueatum,** Hall? (Pal. New York, t. 56, fig. 1). I doubt this reference. McCoy refers to *O. subcostatum*, Portlock, as a near ally.	a. 629, Dinas Bran, Llangollen.
G d 8	*Cycloceras*, p. 320.	**Orthoceras subannulare,** Münst. (Beiträge, Heft III. p. 99, pl. 19, fig. 3). See above in Lower Wenlock, p. 98.	b. 829, Builth Bridge, Wenlock Shale.
G d 8	*O. bullatum*, p. 313.	**Orthoceras,** sp. Much finer striæ than *O. bullatum*, and more like *O. tenuistriatum*, perhaps identical.	a. 635, Llyn Alwen, Denbighshire.
G d 8	p. 316.	**Orthoceras primævum,** Forbes (*Creseis*, Forbes, Quart. Journ. Geol. Soc. Vol. II. pl. 18, fig. 2). Long thin shells with very convex septa, easily crushed.	b. 833, Cwmbach Builth; a. 636, Builth Bridge.
G d 8	p. 318.	**Orthoceras ventricosum,** Sharpe (*Creseis*, Quart. Journ. Geol. Soc. Vol. II. pl. 13, fig. 3). A shorter shape than *O. primævum*.	b. 830, Builth Bridge (W. Shale).
G d 10		**Orthoceras,** sp. New, large like *O. filosum*, but with impressed lines instead of raised thread-like striæ. It is probably related to *O. dulce*, Barr.	a. 638, Dudley, F. C.
F C		**Orthoceras distans,** Sow.? (Siluria, 2nd ed. pl. 26, fig. 4), has in this specimen fine longitudinal striæ, not before observed. (It is imperfect, and may not be certainly Sowerby's species. I think it is.)	a. 619, Dudley, F. C.
G d 10	p. 313.	**Orthoceras angulatum,** Wahl.? (Siluria, 2nd ed. pl. 28, fig. 4). These specimens (if the same as the well-known casts in Ludlow rocks) shew the ribs much more raised and thread-like—we had only casts before.	a. 643, a. 645, a. 644 may be quite distinct, I think it is so. Dudley, F.C.
		**Phragmoceras,** Brod. A laterally compressed and curved shell, aperture contracted in the middle, siphuncle ventral, radiated.	

Case and Column of Drawers.	Reference to McCoy's Synopsis: and Figures of Genera.	Names and References; Observations, &c.	Numbers and Localities.
F C	p. 322.	**Phragmoceras ventricosum,** Sow. (Siluria, 2nd ed. pl. 32).	a. 616, Dudley, F. C.
F C		**Phragmoceras ventricosum,** var. dwarf with close septa. A little like *P. arcuatum,* Sow.	a. 617, Dudley, F. C.
F C	*Poterioceras,* p. 321. *P. ellipticum,* p. 321. 	**Phragmoceras (Gomphoceras) pyriforme,** Sow. (Siluria, 2nd ed. pl. 30, figs. 1—3).	a. 622, Dudley, F. C.
F C G d 10 G d 8		**Phragmoceras (Gomphoceras) æquale,** Salter.	a. 623, Dudley, F. C.; b. 835, Tortworth (Earl Ducie); b. 834, Dudley, F. C.
		Lituites, Breyn. [The genus *Lituites* differs from *Trochoceras* in being discoid and not spiral, and from *Hortolus* by the whorls of the spire being in contact.]	Halfway House, on the road from Llandovery to Trecastle.
G d 10		**Cyrtoceras Biddulphii,** Sow. (*Lituites,* Siluria, 2nd ed. pl. 31, figs. 1, 2).	a. 637, Dudley, F. C.
		[**Cyrtoceras? compressum,** Sow. (*Phragmoceras,* Siluria, 2nd ed. pl. 31, fig. 4.)]	
F C		**Cyrtoceras corniculum,** Barrande (Sil. Syst. Bohem. Vol. II. p. 492, pl. 121); perhaps a *Phragmoceras,* section round.	a. 914, Dudley.
		**Trochoceras** (Barr.), formerly known as *Lituites,* but found to be spiral, not discoid shells.	
G b		**Trochoceras spurium,** Salter, n. sp. Much narrower whorls than *Phragm. nautileum.*	a. 466, Wenlock Shale of Builth Bridge.
F C		**Trochoceras,** sp. Narrow whorls, very faint ribs.	a. 913, Dudley, F. C.
G d 8 G d 10	*Hortolus,* p. 324.	**Trochoceras giganteum,** Sow. (*Lituites,* Siluria, 2nd ed. pl. 36, figs. 1—3). Very common, but seldom perfect. The species vary as *Ammonites* and *Nautili* do, and *Trochoceras* represents, among the Nautilidæ, *Helicoceras,* D'Orb., among the Ammonitidæ.	a. 952, Ledbury; a. 641, a. 642, Dudley, F. C.

LUDLOW ROCKS.—Lower Division.

Not easily distinguishable, except by their superposition and fossils, from the Wenlock rocks. The mineral character, shale and limestone, is much, of course, like that of the underlying Wenlock limestone and shale. And of course, also, many of the organic remains are alike. A great many of them are corals, bryozoa, and brachiopod and cephalopod shells, which have considerable range. But there is no practical difficulty at all in recognizing Ludlow from Wenlock by the fossils. It is very easy; and it is the business of the scientific geologist to neglect the common terms, and study the real and abiding differences which constitute the claim of one set of rocks to be called distinct from another. Moreover, the Ludlow is distinct from the Wenlock in Sweden; and the Lower Helderberg series of America (N.) are the exact equivalents of our Lower and Upper Ludlow rocks, as the Niagara group is of the Wenlock formation.

I include the Aymestry with the Lower Ludlow, as one is only a calcareous condition of the other. And Mr Lightbody of Ludlow has already laid stress on the fact (Quart. Geol. Journ. Vol. 19, p. 368) that the calcareous nature of the beds above the Aymestry rock influences the fossils to a marked degree; so that, were it possible, it would be better to draw the line "Upper Ludlow" only at some distance above the Aymestry rock. Even in districts where the Aymestry is absent, the Lower and Upper Ludlow rocks are recognizable; though not easily separable by a hard line. (J. W. Salter, from personal survey.)

LOWER LUDLOW ROCK AND AYMESTRY LIMESTONE.

Case and Column of Drawers.	Reference to McCOY's Synopsis: and Figures of Genera.	Names and References; Observations, &c.	Numbers and Localities.
		PLANTÆ.	
G e 1	*Sponges.*	**Spongarium Edwardsii,** Murch. (Siluria, 2nd ed. pl. 12, fig. 3).	b. 2, Dinas Bran, Llangollen.
		AMORPHOZOA.	
		? New flat sponge with radiating fibres. Shape cylindro-conical, two inches long.	See Bryozoa.
		HYDROZOA (BRYOZOA or POLYZOA.)	
G e 1	*G. ludensis,* p. 4.	**Graptolithus priodon,** Bronn. (Leth. Geogn. t. 1, fig. 13).	b. 850, Garden Quarry, Aymestry.
G e 1		**Graptolithus priodon, G. ludensis,** var. minor, McCoy, p. 5.	b. 851, Leintwardine, Shropshire.
G e 1		**Graptolithus,** sp.	b. 852, Underbarrow, Kendal.
G e 1		**Dendrograpsus,** Hall.	
G e 1		**Dendrograpsus,** sp. Nicholson, Quart. Journ. Geol. Soc.	b. 853, Leintwardine.

Case and Column of Drawers.	Reference to McCoy's Synopsis: and Figures of Genera.	Names and References; Observations, &c.	Numbers and Localities.
G e 1		**New Bryozoon?** or Sponge.	a. 520, Leintwardine.
		ZOOPHYTA, or Corals. (*Cœlenterata.*)	
G e	p. 24.	**Stenopora fibrosa,** Goldf. sp. (*Favosites,* Siluria, 2nd ed. pl. 40, fig. 6).	b. 853*, Sedgley.
G e 1		**Syringopora.**	b. 854, High Thorns, Kendal.
G e 1	Pl. 1 c, fig. 11, p. 36.	**Cyathaxonia Siluriensis,** McCoy.	a. 903, Underbarrow.
		ECHINODERMATA.	
		Actinocrinus, Miller? It is not certain, nor very likely, that the following species belongs to the Carboniferous genus, to which it is assigned. It is quite as likely to be a *Cyathocrinus*, and the ornament is similar.	
G e 1	Appendix A, p. 1, Pl. 1 D, fig. 3.	**Actinocrinus pulcher?** Salter.	a. 904, Shepherd's Quarry, Kendal.
G e 1		**Platycrinus,** Miller.	b. 856, Benson Knott, Kendal.
G e 2		**Platycrinus,** sp.	b. 857, Shepherd's Quarry, Kendal.
G e 2		**Platycrinus,** sp.	a. 519, Shepherd's Quarry, Kendal.
G e 2		**Cyathocrinus,** sp.	b. 855, Shepherd's Quarry, Patton, Kendal.

Case and Column of Drawers.	Reference to McCoy's Synopsis: and Figures of Genera.	Names and References; Observations, &c.	Numbers and Localities.
G e 2	*I. pyriformis*, p. 54. 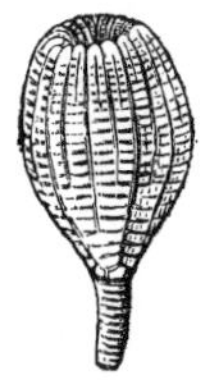	**Ichthyocrinus McCoyanus,** Salter. The branches spread too much, and are too convex in the joints for the Dudley species.	b. 16, Light Beck, Underbarrow.
G e 2	Pl. 1 D, fig. 1, p. 53. 	**Taxocrinus Orbignyi,** McCoy.	a. 518, High Thorns, Underbarrow.
		ASTEROIDEA.	
		**Palasterina,** McCoy, Salter.	
G e 2	*Uraster*, p. 60.	**Palasterina primæva,** Forbes (Siluria, 3rd ed. Foss. 57, fig. 1).	a. 517, High Thorns, Underbarrow, Forbes' fig.
	Uraster, p. 59.	**Palæaster,** Hall, Salter, &c.	
G e G e 2	*Uraster*, p. 59. 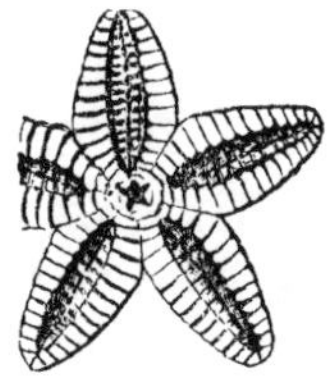	**Palæaster Ruthveni,** Forbes (Siluria, 3rd ed. Foss. 57, fig. 3, Mem. Geol. Surv. Dec. 1, t. 1, fig. 1).	a. 516, a. 920, a. 921, High Thorns, Underbarrow.
G e 2	*Uraster ?* p. 59.	**Palæaster hirudo,** Forbes (Mem. Geol. Surv. Dec. 1, t. 1, fig. 4).	a. 514, Potter Fell, Kendal.
	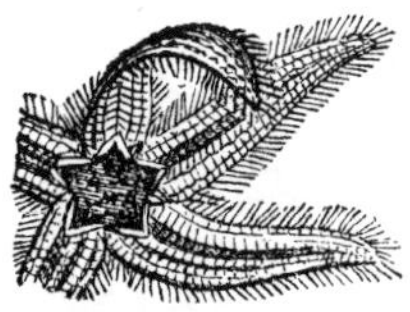	**Palæocoma,** Salter. Divisible into *Palæocoma* proper and *Bdellacoma.*	
G e Ge 2		**Palæocoma Marstoni,** Salter (Siluria, 3rd ed. Foss. 21, fig. 3).	a.515, b.861, Leintwardine.
		Bdellacoma, Salter.	
G e 2		**Bdellacoma vermiformis,** Salter (Ann. Nat. Hist. ser. 2, Vol. 20, p. 329).	b. 858, Leintwardine.

Case and Column of Drawers.	Reference to McCoy's Synopsis: and Figures of Genera.	Names and References; Observations, &c.	Numbers and Localities.
	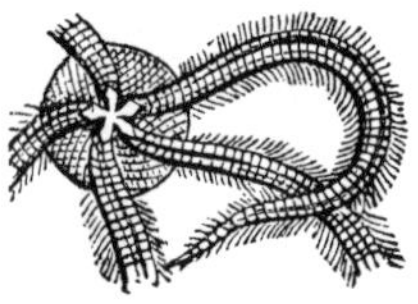	**Protaster,** Forbes. (Also Salter, Ann. Nat. Hist. 2nd Ser. Vol. 20.)	
G e 2 G e		**Protaster Miltoni,** Salter (Siluria, 3rd ed. Foss. 21, figs. 1, 2).	b. 859, b. 860, Leintwardine.
G e	p. 60.	**Protaster Sedgwicki,** Forbes (Mem. Geol. Surv. Decade 1, Pl. 4).	a. 919, Docker Park, Kendal.
		ANNELIDA.	
G e 3	Appendix A, p. 1, Pl. 1 D, figs. 11, 12.	**Serpulites dispar?,** Salter.	b. 862, Tenterfell, Kendal, N. end.
G e 7		**Sabella? Murchisoniana,** n. sp.	b. 1, Dudley, on Orthoceras, F. C.
G e 3	p. 131. 	**Spirorbis tenuis?** Sow. (*S. Lewisii*, Siluria, 3rd ed. pl. 16, fig. 2).	b. 863, Green Quarry, Leintwardine.
		CRUSTACEA. Phyllopoda.	
	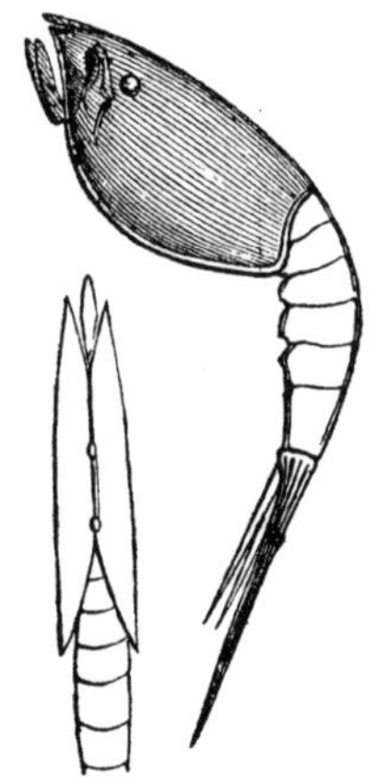	**Ceratiocaris,** McCoy. "Carapace bi-"valve, the dorsal line simply an-"gulated (? undivided), with a slight "furrow beneath it on each side; "sides semi-elliptical, much elon-"gated from before backwards, "evenly convex; ventral margin "gently convex, posterior end ab-"ruptly truncate obliquely." Synopsis, p. 136.	
G e	Pl. 1 E, figs. 7, 7a, *Pterygotus lept.*, p. 175.	**Ceratiocaris leptodactylus,** McCoy.	a. 923, a. 924, Leintwardine.
G e	Same tablet, figs. 7c, 7d.	**Ceratiocaris robustus,** Salter (Ann. Nat. Hist. ser. 3, Vol. 5, p. 158).	a. 925, Leintwardine.

Case and Column of Drawers.	Reference to McCoy's Synopsis: and Figures of Genera.	Names and References; Observations, &c.	Numbers and Localities.
	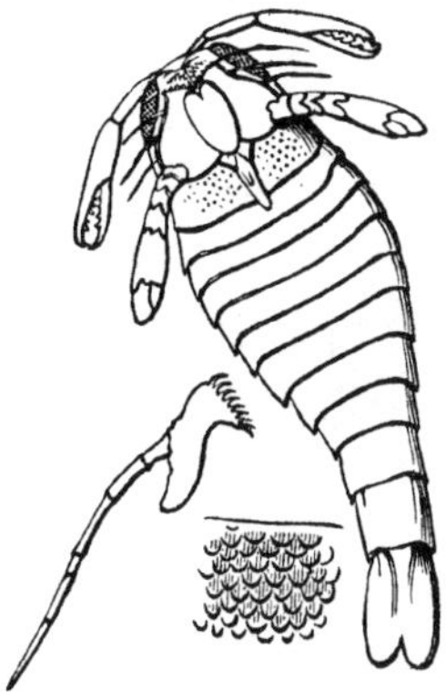	**CRUSTACEA,** Merostomata, Sub-Order Eurypterida. **Pterygotus,** (Agassiz), Huxley, Salter and Woodward. The giant of the *Crustacea,* attaining a size of seven to eight feet in length, and abundant in species. At first regarded as parts of a fish, it was soon restored to its true position by the renowned Agassiz, and it has since received attention from the above authors. (See Upper Ludlow.)	
Ge 3		**Pterygotus?** A thick anterior joint of some large species. (It can scarcely be the cup of a large crinoid, much obliterated.)	a. 521, Underbarrow, Kendal.
Ge		**Pterygotus arcuatus,** Salter (Mem. Geol. Surv. Mon. 1, on *Pterygotus,* p. 95, t. 13). The difference in sculpture and shape of the antennæ marks this species from *P. anglicus* and its allies.	b. 864, Leintwardine.
	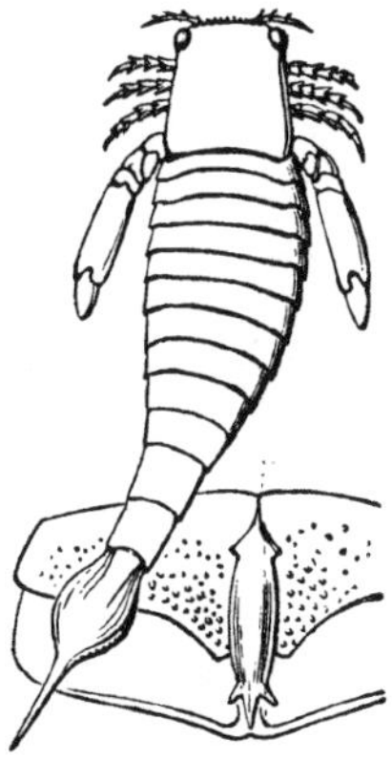	**Slimonia** (Page, H. Woodward). [Slimonia differs from Pterygotus in shape and in the absence of the great chelate antennæ. The thoracic plate covering the sexual organs differs in male and female.]	
Ge		**Slimonia punctata** (*Pterygotus,* Salter, Mem. Geol. Surv. Mon. 1, pl. 10, 11). A body-ring only.	b. 865, Leintwardine.
		CRUSTACEA, Trilobita, Burm.	
Ge 3	*Forbesia,* p. 174. 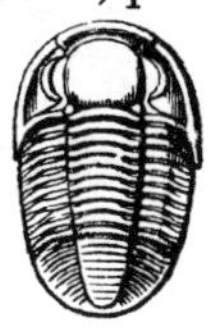	**Proetus latifrons,** McCoy, sp. (Silur. Foss. Ireland, t. 4, fig. 11, Mem. Geol. Surv. Vol. 2, pt. 1, t. 6, fig. 1).	a. 905, Underbarrow.

Case and Column of Drawers.	Reference to McCoy's Synopsis; and Figures of Genera.	Names and References; Observations, &c.	Numbers and Localities.
G e 3	*Odontochile caudata*, v. *minor*? p. 160.	**Phacops Downingiæ,** Murch. A large specimen of a common species. It belongs to the section *Acaste*.	a. 522, Underbarrow.
G e 3	*Portlockia*, p. 163. 	**Phacops** (Portlockia) **Stokesii,** M. Edw. (Siluria, 3rd ed. pl. 10, fig. 6; pl. 18, fig. 6).	b. 866, Garden Quarry, Aymestry.
G e 3	*C. subdiademata*? p. 166.	**Calymene Blumenbachii,** Brong.	b. 867, Aymestry.
G e	p. 167.	**Calymene tuberculosa,** Salter (Mem. Geol. Surv. Vol. 2, pt. 1, p. 342, t. 12). Has a much longer snout in front, and more depressed and small glabella than the preceding.	a. 927, High Thorns, Underbarrow.
G e		**Acidaspis coronatus,** Salter (Quart. Journ. Geol. Soc. Vol. 13, p. 210. *A. Brightii*, Salt. Mem. Geol. Surv. Vol. 2, pt. 1, pl. 9, figs. 8, 9).	a. 658, Dudley Shales above limestones, F. C.
		BRACHIOPODA.	
G e 3	 p. 190.	**Discina Morrisii,** Davidson (Sil. Brach. p. 65, t. 7, figs. 10—12).	b. 868, Leintwardine.
G e	*L. cornea*, p. 251. 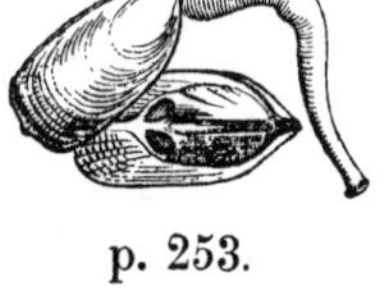	**Lingula minima,** Sow. (Sil. Syst. t. 4, fig. 49).	b. 869, Mortimer's Cross.
G e 3		**Lingula lata,** Sow. (Siluria, 3rd ed. pl. 20, fig. 6).	b. 870, Leintwardine.
G e	p. 253.	**Lingula Lewisii,** Sow. (Siluria, 3rd ed. pl. 20, fig. 5).	b. 871, Sedgley.
G e 4		**Spirifer** (Cyrtia), *trapezoidalis*, Dalm., (*S. exporrectus*, Wahl. Siluria, pl. 9, fig. 24).	b. 872, Woolhope.

Case and Column of Drawers.	Reference to McCoy's Synopsis: and Figures of Genera.	Names and References; Observations, &c.	Numbers and Localities.
G e 4		New genus, or *Athyris?*	a. 525, Llangammarch, Sugar Loaf, Llandovery.
G e 4	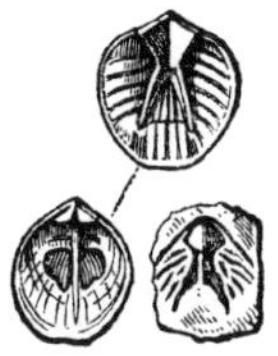	**Meristella?** sp. A lobed shell with only lines of growth.	b. 873, Llyn Alwen, Denbighshire.
G e 5 G e 4	*Hemithyris*, p. 201.	**Meristella didyma,** Dalm., sp. (Siluria, 3rd ed. pl. 22, fig. 15).	a. 937, b. 875, Sedgley, ventral valve interior, F. C.
G e 4	*Siphonotreta anglica!* p. 188.	**Athyris? obovata,** Sow. (Siluria, 3rd ed. pl. 22, fig. 16, Davidson, Sil. Brach. p. 121, t. 12, fig. 19).	a. 524, Sunny Banks, Grayrigg, Westmorland; b. 874, Dudley, F. C.
G e 4 G e 5	*Hemithyris*, p. 204. 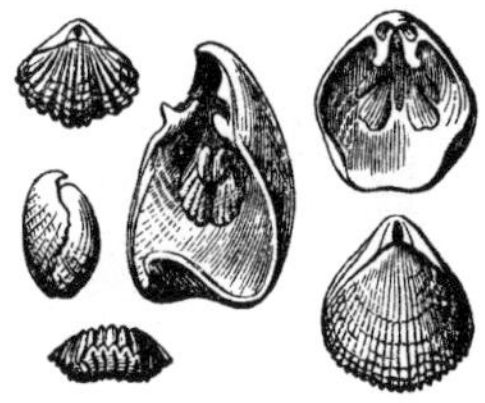	**Rhynchonella nucula,** Sow. (Siluria, 3rd ed. pl. 9, figs. 9, 11; pl. 22, figs. 1, 2).	b. 876, Leintwardine; b. 877, Cwm Craig ddu; a. 938, Sedgley, F. C.
G e 5 G e G e 4	*Hemithyris*, p. 204.	**Rhynchonella navicula,** Sow. sp. (Siluria, 3rd ed. pl. 22, fig. 12, Davidson, Sil. Brach. p. 190).	b. 878, Erw Gilfach, Builth; a. 939, a. 940, Sedgley, F.C.; b. 879, N. end of Potter Fell, Kendal; b. 880, Cowan Head, Kendal; Cwm Craig ddu, Builth; Mynydd-y-gaer, S. side; Leintwardine.
G e 4 G e G e 5	*Hemithyris sphæroidalis*, pl. 1 L, fig. 4, p. 206.	**Rhynchonella Wilsoni,** Sow. sp. var. (Davidson, Sil. Brach. p. 173, t. 23, fig. 10).	a. 941, a. 942, Sedgley, var. *vera*, F. C.; a. 523, Botville, Church Stretton; b. 881, Sedgley.
G e 4	*Hemithyris*, p. 200.	**Rhynchonella crispata,** Sow. var.	b. 882, Leintwardine.
G e 4	p. 209.	**Pentamerus Knightii,** Sow. var. *elongatus*, McCoy.	b. 884, Woolhope.

Case and Column of Drawers.	Reference to McCoy's Synopsis: and Figures of Genera.	Names and References; Observations, &c.	Numbers and Localities.
Ge 5 Ge 4 Ge	p. 209.	**Pentamerus Knightii,** Sowerby (Min. Conch. t. 28, Davidson, Sil. Brach. t. 16, 17).	b. 883, Woolhope ridges; b. 885, Walsall, F. C.; a. 933, Aymestry, interior; b. 886, Sedgley.
Ge		**Pentamerus Knightii,** var. **Aylesfordii,** Sow. (Min. Conch. t. 29).	a. 934, Aymestry.
Ge 4		**Pentamerus galeatus,** var. The interior of the ventral valve has been sometimes taken for the very different species *P. globosus.*	b. 887, Leintwardine; b. 888, above Park Lane, Llandeilo.
Ge 4 Ge	*Spirigerina,* p. 198. 	**Atrypa reticularis,** Linn. sp. (Siluria, 3rd ed. pl. 9, fig. 1; pl. 21, figs. 12, 13; Davidson, Sil. Brach. p. 129, pl. 14, figs. 1—22).	b. 889, Brockton and Burton; b. 890, Leintwardine; b. 891, Park Lane, Llandeilo; b. 892, Collinfield, Kendal.
Ge 4	p. 216.	**Orthis elegantula,** Dalm. (Siluria, 3rd ed. pl. 5, fig. 5).	b. 893, Mynydd-y-gaer, Llanefydd.
Ge 5	*O. lunata,* p. 220.	**Orthis orbicularis,** Sow. (Siluria, pl. 20, fig. 9).	a. 935, a. 936, Sedgley, F. C.
Ge 4	*Leptæna,* p. 244.	**Strophomena funiculata,** McCoy, sp (Sil. Foss. Ireland, t. 3, fig. 11).	b. 894, Park Lane, Llandeilo.
Ge 4	*Leptæna,* p. 243. 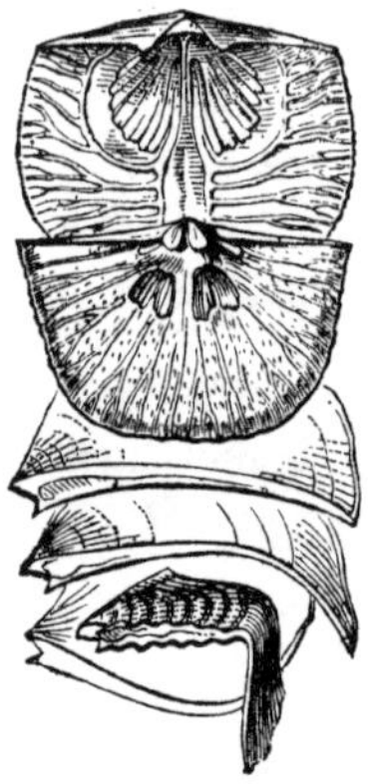	**Strophomena filosa,** Sow. sp. (*Orthis,* Siluria, 3rd ed. pl. 20, fig. 21).	b. 895, Tullithwaite Hall, Underbarrow. b. 896, Sedgley.

Case and Column of Drawers.	Reference to McCoy's Synopsis: and Figures of Genera.	Names and References; Observations, &c.	Numbers and Localities.
G e 4	*Leptæna*, p. 243.	**Strophomena euglypha,** Dalm. sp. (Siluria, 3rd ed. pl. 20, fig. 19).	b. 897, Leintwardine.
G e 4	*Leptæna*, p. 235.	**Chonetes minima,** Sow. (Siluria, 3rd ed. pl. 20, fig. 16).	b. 898, above Rother Bridge.
G e 5	*Chonetes*, p. 249.	**Chonetes lata,** Von Buch. (*C. striatella*, Dalm. sp. Davidson, Sil. Brach. p. 331, t. 49, figs. 23—26).	b. 900, Sedgley, F. C.
		LAMELLIBRANCHIATA.	
		Pseudaxinus, Salter.	
G e 5	*Anodontopsis*, p. 272. Pl. 1 L, fig. 9.	**Pseudaxinus securiformis,** McCoy.	b. 899, Dudley, F. C.
		Anodontopsis, McCoy. A genus of thin shells, allied to *Modiolopsis*, but with very much enlarged posterior area. (The following may not belong to the genus.)	
G e 5	Pl. 1 K, fig. 10, p. 272.	**Anodontopsis quadratus,** Salter, sp. (Mem. Geol. Surv. Vol. II. Pt. 1, p. 363, pl. 20, fig. 1).	a. 944, a. 945, Dudley, F. C.
G e G e 5	p. 263. 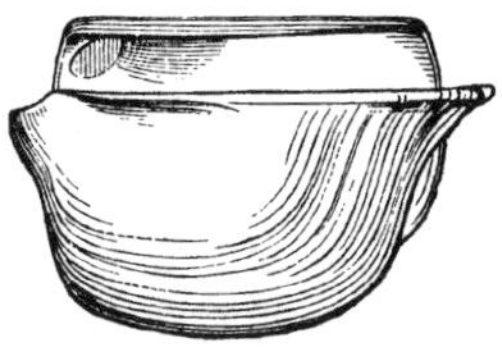	**Pterinea Sowerbyi,** McCoy (Siluria, 2nd ed. pl. 23, fig. 15).	a. 808, Dudley, F. C.; a. 929, Leintwardine.
G e 6	Pl. 1 I, figs. 1, 2, p. 261.	**Pterinea pleuroptera,** Conrad. sp.	a. 952, Park Lane, Llandeilo.
G e 5		**Pterinea** Condor, n. sp. Very wide hinge-line three inches broad.	a. 809, a. 810, Dudley, F. C.
G e 5	 p. 262.	**Pterinea lineatula,** D'Orb. (*Avicula lineata*, Sow. Sil. Syst.; Siluria, 2nd ed. pl. 23, fig. 16).	b. 728, Dudley, F. C.
G e 6		**Pterinea retroflexa,** Wahl. sp. (Leth. Suec. t. 17, fig. 12; Siluria, pl. 9, fig. 26).	b. 910, High Thorns, Underbarrow; b. 911, Leintwardine.
G e 6	*Pterinea hians*, p. 260.	**Pterinea retroflexa,** var. *hians*.	a. 950, Mortimer's Cross.
G e 6	Pl. 1 I, fig. 6.	**Pterinea,** sp. like *Sowerbyi*.	a. 948, Dudley.

Case and Column of Drawers.	Reference to McCoy's Synopsis: and Figures of Genera.	Names and References; Observations, &c.	Numbers and Localities.
G e 6	Pl. 1 I, fig. 4, p. 263.	**Pterinea tenuistriata,** McCoy.	a. 951, Cwm Craig ddu; b. 912, Park Lane, Llandeilo.
G e 5	Pl. 1 L, fig. 21. *M. Nillsoni,* p. 267.	**Modiolopsis gradatus?** Salter (Mem. Geol. Surv. Vol. II. pt. 1, p. 363, pl. 20, figs. 3—5.)	a. 943, Dudley, F. C.
G e 5		**Modiolopsis,** sp. Small transverse sp.	b. 901, Dudley, F. C.
	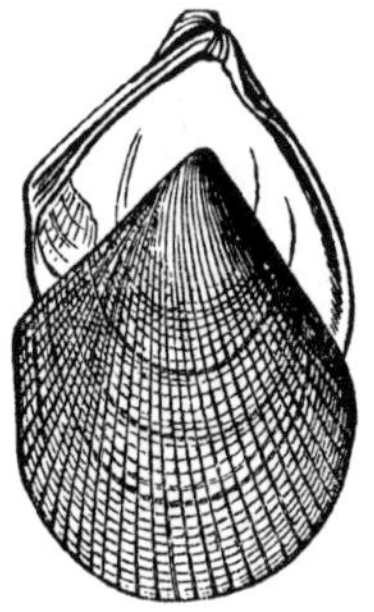	**Ambonychia,** Hall, Pal. N. York, Vol. I. [Differs from *Pterinea* and *Avicula* by the short hinge-line and no anterior wing, and from *Inoceramus* by the absence of transverse pits in the hinge.]	
G e 6 G e	p. 264.	**Ambonychia striata,** Sow. sp. allied to *Lunulucardium* (*Cardium,* Sil. Syst. t. 6, fig. 2).	b. 902, above Park Lane, Llandeilo; b. 903, Garden Quarry, Aymestry.
G e 6	Pl. 1 K, fig. 16, p. 264.	**Ambonychia acuticostata,** McCoy.	a. 949, Dinas Bran, Llangollen, or Wenlock Shale.
G e 6 G e	p. 282. 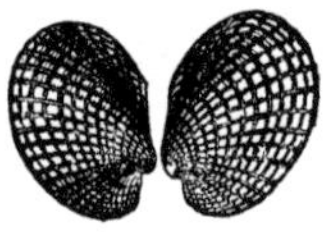	**Cardiola interrupta,** Brod. (Siluria, 2nd ed. pl. 23, fig. 12).	b. 904, Leintwardine; b. 905, Dudley; b. 906, Yr-Allt, Welchpool; b. 907, Cwm Craig ddu, Builth; b. 908, Mynydd-y-gaer; b. 909, Sugar Loaf, Llandovery.
G e 5	[Not of McCoy, p. 275.] 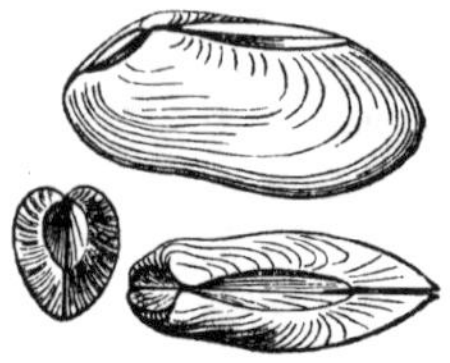	**Orthonotus semisulcatus,** Sow. sp. (*Modiola,* Sil. Syst. p. 617, t. 8, fig. 6).	a. 946, Dudley, F. C.
G e 5		**Orthonotus undatus?** Sow. (Siluria, pl. 23, fig. 4).	a. 947, Dudley, F. C.
G e 6		**Cuculella coarctata,** Phill. (Mem. Geol. Surv. Vol. II. pt. 1, p. 366).	b. 913, Dinas Bran.
G e 6	*Nucula,* p. 285. 	**Ctenodonta anglica,** D'Orb. (Siluria, 3rd ed. pl. 23, fig. 10).	b. 914, High Thorns, Underbarrow.

Case and Column of Drawers.	Reference to McCoy's Synopsis: and Figures of Genera.	Names and References; Observations, &c.	Numbers and Localities.
		**PTEROPODA.**	
Ge 8		**Graptotheca catenulata,** n. sp. Salter.	a. 986, Leintwardine.
		Theca.	
Ge 8	p. 287. 	**Conularia Sowerbyi,** Defr. (Siluria, 3rd ed. pl. 25, fig. 10. *C. quadrisulcata,* Sil. Syst. t. 12, fig. 22).	b. 915, Park Lane, Llandeilo; b. 916, High Thorns, Underbarrow.
Ge	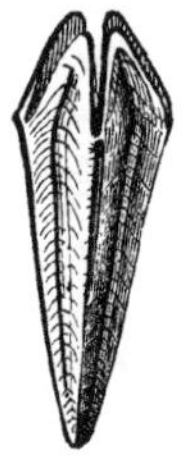	**Conularia bifasciata,** Salter MSS. One specimen only shews one face, and that of a young specimen.	a. 926, Leintwardine.
		GASTEROPODA.	
Ge 8	1 K, fig. 39, p. 290. 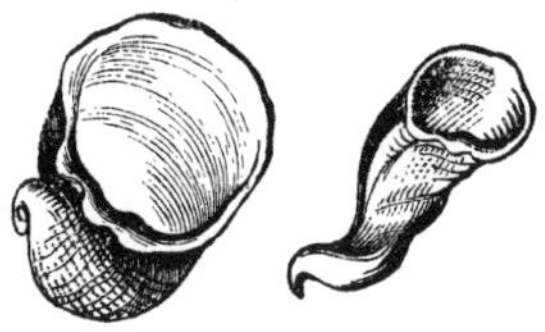	**Acroculia euomphaloides,** McCoy.	a. 624, Green Quarry, Leintwardine.
Ge	p. 292. 	**Pleurotomaria undata,** Sow. (Siluria, 2nd ed. pl. 24, fig. 6).	a. 922, Green Quarry, Leintwardine.
Ge 7 Ge 8		**Pleurotomaria striatissima,** Salter.	a. 987, Dudley, F. C.; a. 991, Green Quarry, Leintwardine.
Ge Ge 7	p. 293.	**Murchisonia Lloydii,** Sow. (Siluria, 2nd ed. pl. 24, fig. 5).	b. 917, Garden Quarry, Aymestry; b. 918, Dudley, F. C.

Case and Column of Drawers.	Reference to McCoy's Synopsis: and Figures of Genera.	Names and References; Observations, &c.	Numbers and Localities.
		Section *Hormotoma*.	
G e	*Murchisonia*, p. 293.	**Murchisonia cingulata,** Hisinger (Leth. Suec. t. 12, fig. 6).	b. 919, Leintwardine.
G e 7		**Murchisonia articulata,** Sow. (Siluria, 2nd ed. pl. 24, fig. 2).	b. 920, Dudley, F. C.
G e 8	*Litorina*, p. 305.	**Cyclonema octavia,** D'Orb. (Siluria, 3rd ed. pl. 24, fig. 4).	b. 921, High Thorns, Underbarrow, Kendal; b. 922, Mortimer's Cross.
G e 8	*Litorina*, p. 305.	**Cyclonema corallii,** Sow. (Siluria, 3rd ed. pl. 24, fig. 1).	b. 923, Sedgley.
G e 8	Pl. 1 K, fig. 46, p. 306.	**Cyclonema undifera,** McCoy.	a. 918, Mortimer's Cross.
G e 8	p. 297.	**Euomphalus centrifugus,** Wahl. sp. (His. Leth. Suec. t. 12, fig. 1).	b. 925, Leintwardine.
G e 8	Pl. 1 K, fig. 35, p. 302.	**Naticopsis glaucinoides,** Sow. (Sil. Syst. t. 3, fig. 14).	b. 924, Benson Knott, Kendal.
G e 8	Pl. 1 K, fig. 33, p. 303.	**Holopella gracilior,** McCoy.	a. 988, Dinas Bran.
G e 8	Pl. 1 L, fig. 16, p. 304.	**Holopella intermedia,** McCoy.	a. 990, High Thorns, Underbarrow.
G e 7		**Loxonema sinuosa,** Sow. (Siluria, 3rd ed. pl. 24, fig. 3).	b. 926, Dudley, F. C.
G e 8	Pl. 1 K, fig. 34, p. 302.	**Loxonema elegans,** McCoy.	a. 989, Green Quarry, Leintwardine.

Case and Column of Drawers.	Reference to McCoy's Synopsis: and Figures of Genera.	Names and References; Observations, &c.	Numbers and Localities.
		HETEROPODA.	
G e 7		**Bellerophon,** n. sp. More prominent keel than *B. dilatatus.*	b. 927, Dudley, F. C.
G e 7		**Cyrtolites lævis,** Sow. sp. (Silur. Syst. t. 8, fig. 21).	b. 936, Dudley, F. C.
		CEPHALOPODA.	
G e 8	p. 314.	**Orthoceras filosum,** Sow. (Siluria, 2nd ed. Foss. 61, pl. 27, fig. 1). Large and small threads or thin ridges.	b. 928, Garden Quarry, Aymestry.
G e	do.	**Orthoceras dimidiatum,** Sow., var. with distant ribs or plications, which occur only half across the shell—ventral or dorsal aspect, but we know not which.	a. 928, Leintwardine.
G e 7	*Cycloceras,* p. 319.	**Orthoceras Ibex,** Sow. (Siluria, 3rd ed. pl. 29, figs. 3, 4).	b. 929, Dudley, F. C.
G e 8	Pl. 1 L, fig. 31, p. 320. *Cycloceras.*	**Orthoceras tenui-annulatum,** McCoy. Surely a variety of *O. Ibex.* It has only fine striæ.	a. 992, Green Quarry, Leintwardine; a. 995, near Aymestry.
G e 7		**Orthoceras,** sp. like *Ibex,* but much larger. See *Sabella* in the annelid drawer.	a. 983, Dudley, F. C.
G e 7	p. 313.	**Orthoceras angulatum,** Wahl. (Siluria, 2nd ed. pl. 28, fig. 4). Has a set of parallel lines between the ribs.	a. 984, Dudley (striæ good), F. C.
G e 8	p. 317.	**Orthoceras subundulatum,** Portl. (Geol. Report, t. 28, fig. 2).	b. 930, Garden Quarry, Aymestry.
G e 8		**Orthoceras subundulatum?** Portl.	b. 931, Clungunford.
G e G e 8	*Hortolus Ibex,* p. 324. *Lituites articulatus,* p. 323.	**Orthoceras perelegans,** Salt. (Mem. Geol. Surv. Vol. II. pt. 1, pl. 13, figs. 2—4, p. 354).	b. 932, Garden Quarry, Aymestry; b. 933, High Thorns, Underbarrow.

Case and Column of Drawers.	Reference to McCoy's Synopsis: and Figures of Genera.	Names and References; Observations, &c.	Numbers and Localities.
		Phragmoceras, Broderip.	
	p. 322.	**Phragmoceras ventricosum,** Sow. (Siluria, 2nd ed. pl. 32), is so common in the Lower Ludlow, that we insert it.	
G e	do. 	**Phragmoceras? intermedium,** McCoy (Siluria, 2nd ed. pl. 30, fig. 4).	a. 930, Leintwardine.
G e	do.	**Phragmoceras (Gomphoceras) pyriforme,** Sow. (Sil. 2nd ed. pl. 30, figs. 1—3).	b. 934, near Aymestry. Leintwardine, Dudley, F. C.
G e 8	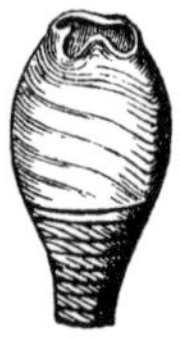	**Phragmoceras (Gomphoceras) liratum,** n. s. Salter. An open wide-mouthed species—(if of this genus)—it has not attained to the contracted portion of the mouth.	a. 993, Garden Quarry, Aymestry.
G e 8	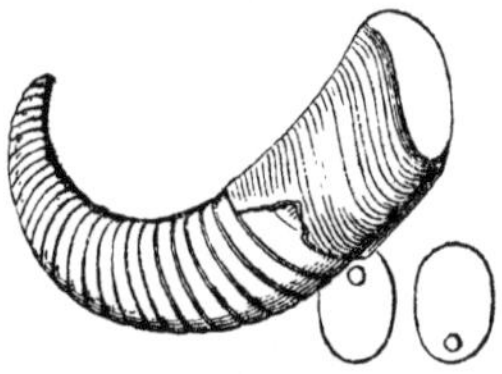	**Cyrtoceras,** sp.	a. 994, Llangammarch, Sugar Loaf, Abergavenny.
		**Trochoceras,** Barrande, Syst. Silur. de la Bohêm. Vol. II. p. 74.	
G e 8		**Trochoceras,** sp. Distorted fragment.	b. 935, Dudley.
G e	*Hortolus,* p. 324.	**Trochoceras giganteum,** Sow. sp. (*Lituites,* Siluria, 2nd ed. pl. 33, figs. 1—3).	a. 931, a. 932, Leintwardine, straight portion.
G e 7	p. 323.	**Lituites articulatus,** Sow. (Sil. 2nd ed. pl. 31, fig. 6). Three species formerly went under this name. But the flat discoid and keeled shell like a small Lias ammonite is the one to which the name must be restricted.	a. 985, Dudley, F. C.

Upper Ludlow Rocks.

1. *Mudstones*—with calcareous bands, especially near the top.
2. *Bone-bed*—an inch to a foot in thickness.
3. *Downton Sandstone.* Of the Silurian district; represented by a red rock at Usk, and by "tile-stones" in the S. Welch district.

(The Bone-bed and Downton Sandstone are separately catalogued: but it is not always easy, in the absence of the very persistent Bone-bed, to draw the line between them.)

Case and Column of Drawers.	Reference to McCoy's Synopsis: and Figures of Genera.	Names and References; Observations, &c.	Numbers and Localities.
		PLANTÆ.	
		To the group of *Nullipores* (and especially to the genus *Acetabularia*) are referred several thin calcareous expansions which occur in Ludlow rocks, but are particularly characteristic of Upper Ludlow. True sea-weeds are also found, but rarely. The want of calcareous matter in the fossils (originally abundant) is easily explained, as they occur in sandy rock.	
	p. 42.	[**Spongarium Edwardsi,** Murch. Sil. Syst. t. 26, fig. 12. Siluria, 2nd ed. pl. 12, fig. 3. For the type, see p. 42.]	
G f G e 9	Pl. 1 B, fig. 14, p. 43.	**Spongarium interlineatum,** McCoy. Oval, quite regular, a very beautiful fossil. A strong central depression with fine radii, and concentric *rugæ*, very regular.	a. 997, Brigsteer, Kendal; a. 998, b. 56, Benson Knott, Kendal.
G e 9	Pl. 1 B, fig. 15, p. 42.	**Spongarium æquistriatum,** McCoy. Oval and regular, but with coarse *rugæ* and sharp coarse radii.	a. 996, Benson Knott.
G e 9	Pl. 1 B, figs. 16, 17, p. 43.	**Spongarium interruptum,** McCoy. Ridges alternately large and small, interrupted and irregularly prominent.	a. 999, Spital, Kendal.
	Sponges.	**AMORPHOZOA, Sponges.**	
G f	Pl. 1 D, fig. 9.	**Pasceolus Goughii,** Salter, MSS. The genus has been described by Billings from Canada.	b. 53, b. 54, base, Benson Knott.

Case and Column of Drawers.	Reference to McCoy's Synopsis: and Figures of Genera.	Names and References; Observations, &c.	Numbers and Localities.
G f	Pl. 1 D, fig. 7, undescribed. 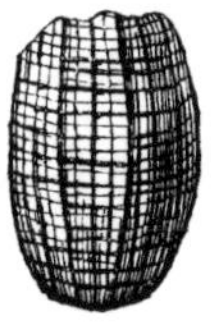	**Tetragonis Danbyi,** McCoy. Oval with rectangular areæ produced by the vertical and horizontal fibres.	b. 57, Benson Knott.
		HYDROZOA. **(BRYOZOA or POLYZOA.)**	
G e 9	*G. ludensis,* p. 4.	**Graptolithus priodon,** Bronn (*G. ludensis,* Murch.).	b. 937, Burton and Brockton.
		ZOOPHYTA, (*Cœlenterata*).	
G e 9	p. 24.	**Stenopora fibrosa,** Goldf. (Pet. Germ. t. 28, fig. 3).	b. 938, Woolhope.
G e 9	p. 25.	**Stenopora fibrosa,** var. **regularis,** McCoy.	b. 939, Woolhope; b. 940, Kendal.
G e 9	p. 24.	**Stenopora fibrosa,** var. **lycopodites,** Hall? (Pal. N. Y. Vol. I. pl. 23, figs. 1—3). Probably not the true Bala variety.	b. 941, Woolhope.
G e 9 G f 10 G f	*Nebulipora,* p. 24. Pl. 1 C, fig. 5. 	**Nebulipora** or **Monticulipora papillata,** McCoy. Growing round *Orthoceras,* a common habit of this very pretty species. The bosses are free from cells.	b. 3, Brigsteer; b. 55, b. 155, Sedbergh (Firbank).
		ECHINODERMATA. Few in this formation.	
		ANNELIDA. Tube-covered worms abound in Ludlow rocks.	
		Serpulites, Murch. Easily distinguished by the nacrous shelly cover, with two flattened sides and the edges only thickened. The shell is very often covered with a lace-like network of hexagonal areolæ, apparently ornament.	
G f 10 G e 9	*Trachyderma? lævis,* p. 133. *S. longissimus,* p. 132.	**Serpulites longissimus,** Murch. (Siluria, 2nd ed. pl. 16, fig. 1). A long shelly tube, curved into nearly a circle. *Trach. lævis* is probably from a Ludlow locality and this species.	b. 154, Aymestry, nearly perfect; b. 942, near Ludlow; b. 943, Burton and Brockton.
G e 9 G f 10	Pl. 1 D, figs. 11, 12, p. 132.	**Serpulites dispar,** Salter (Appendix A. page i.). Thin membranous tubes, contrasting strongly with the thick edge.	b. 941, N. of Benson Knott; b. 156, Scalthwaiterigg, N. end of Benson Knott.

Case and Column of Drawers.	Reference to McCoy's Synopsis: and Figures of Genera.	Names and References; Observations, &c.	Numbers and Localities.
G e 9	p. 63.	**Cornulites serpularius,** Schloth. (Siluria, 2nd ed. pl. 10, fig. 2.)	b. 944, Benson Knott; b. 945, Brigsteer, Kendal.
G e 9	p. 64.	**Tentaculites tenuis,** Sow. (Siluria, 2nd ed. pl. 16, fig. 12).	b. 946, Benson Knott; b. 947, Kirkby Moor, Kendal.
	Trilobites.	**CRUSTACEA, Trilobita.**	
G f	p. 168.	**Homalonotus Knightii,** Murch. (Siluria, 2nd ed. pl. 19, figs. 7—9).	b. 951, Tenter Fell, Kendal.
G e 10	*Odontochile caudata,* var. *minor,* ? p. 160.	**Phacops Downingiæ,** Murch. (ib. pl. 18, fig. 2).	b. 4, Benson Knott, Kendal.
		CRUSTACEA, Phyllopoda.	
G g 2	Large and small bivalve Crustacea.	**Dictyocaris Slimoni,** Salter (Ann. Nat. Hist. 1866, p. 162).	Purchased 1869, near Lesmahagow, Lanark, on the Logan Water; b. 128, valves laid open; b. 139, has circular elevations over the whole surface not yet understood.
G e 9	Pl. 1 E, fig. 2, p. 135.	**Beyrichia Klödeni,** McCoy (Jones, Ann. Nat. Hist. 2nd ser. Vol. XVI. pl. 6, figs. 7—9).	b. 948, Cowan Head, Kendal.
G e 9	p. 135.	**Beyrichia Klödeni,** var. **plicata,** McCoy.	b. 950, Underbarrow.
	p. 136.	**Ceratiocaris,** McCoy. Large shrimp-like forms, a broad two-valved carapace, and a long three-pronged tail, and seven body-rings. Some specimens (G. g. 2) shew rostrum in front.	
G e 10		**Ceratiocaris,** teeth of (probably) the following.	b. 11, Beck Mills, Kendal.
G e 10 G g	Pl. 1 E, fig. 4, p. 137.	**Ceratiocaris inornatus,** McCoy. A large species, the striæ very fine and regular. The eye-spot is problematical.	b. 5, b. 6, b. 35, Benson Knott, Kendal.

Case and Column of Drawers.	Reference to McCoy's Synopsis: and Figures of Genera.	Names and References; Observations, &c.	Numbers and Localities.
G e 10 G g	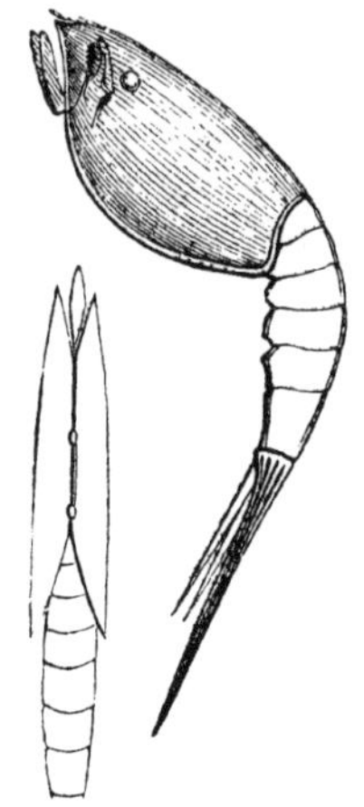	**Ceratiocaris inornatus**, McCoy. Tail spine and body rings!	b. 7, b. 36, Benson Knott.
G e 10 G g	Pl. 1 E, fig. 5, p. 138.	**Ceratiocaris solenoides**, McCoy. A narrow oblong carapace.	b. 8, b. 40, b. 41, Benson Knott.
G g G g 2		**Ceratiocaris papilio**, Salter (Siluria, 2nd ed. Foss. 65, fig. 1).	b. 65, b. 135, with rostrum, Lesmahagow.
G g 2		**Ceratiocaris stygius**, Salter (Ann. Nat. Hist. ser. 3, Vol. v. p. 156). Body and tail and teeth.	b. 136, Lesmahagow.
G e 10	Pl. 1 E, fig. 8, p. 137.	**Ceratiocaris ellipticus**, McCoy. Quite distinct. A small species.	b. 15, Benson Knott.
		CRUSTACEA, Merostomata.	
G g	Pl. 1 E, fig. 12. 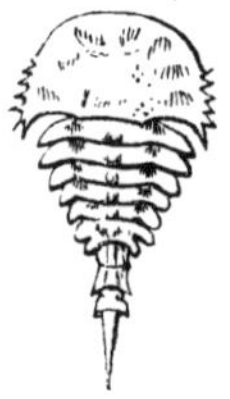	**Hemiaspis aculeatus**, Salter (see Woodward, Quart. Journ. Geol. Soc. Vol. XXI. p. 490).	b. 34, High Thorns, Underbarrow.
G e 10		**Eurypterida (Hemiaspis?)**, striate teeth on margin.	b. 12, Underbarrow.
G g	Pl. 1 E, fig. 21, p. 175.	**Eurypterus Cephalaspis**, Salter (Appendix, A. page v.).	b. 25, Kirkby Moor, Kendal.
G g 2		**Eurypterus lanceolatus**, Salter (Mem. Geol. Surv. Mon. 1, pl. 1, fig. 17).	b. 133, b. 134, Lesmahagow.

Case and Column of Drawers.	Reference to McCoy's Synopsis: and Figures of Genera.	Names and References; Observations, &c.	Numbers and Localities.
G f 10		**Pterygotus,** sp.	b. 151, b. 152, Benson Knott, large species.
G g G g 2 G f 9	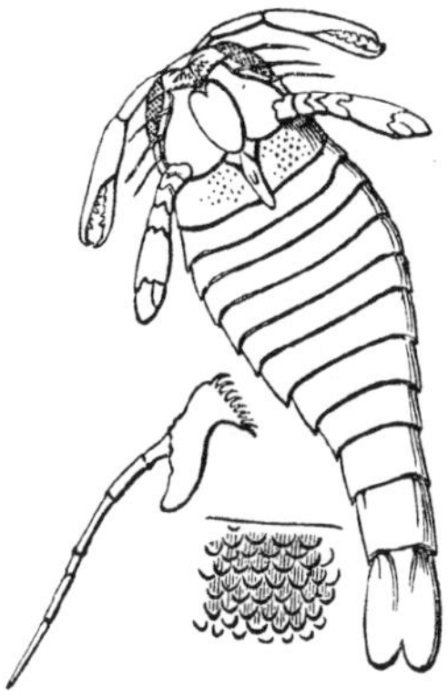	**Pterygotus bilobus,** Salter (Mem. Geol. Surv., Mon. 1, Pterygotus, pl. 1, figs. 1—12, Siluria, 3rd ed. Foss. 26, fig. 1.)	b. 37, shews antennæ and jaws in place; b. 132, antennæ usually reversed; b. 148, b. 149, b. 150, shew the oval metastoma and chelate antennæ; Lesmahagow.
G e 10		**Pterygotus,** sp.	b. 9, Benson Knott.
G e 10		**Pterygotus,** fragment, probably base of swimming foot.	b. 951, Benson Knott.
G f 9		**Pterygotus perornatus,** Salter, (Mem. Geol. Surv. Mon. I. Pterygogus, pl. 1, figs. 13—15.)	b. 147, Lesmahagow.
G e 10		**Crustacean fragment.**	b. 952, Summer How, near Beck Mills, Kendal.
G g G g 1 G g 2 G g 3 G g 4 G f 8	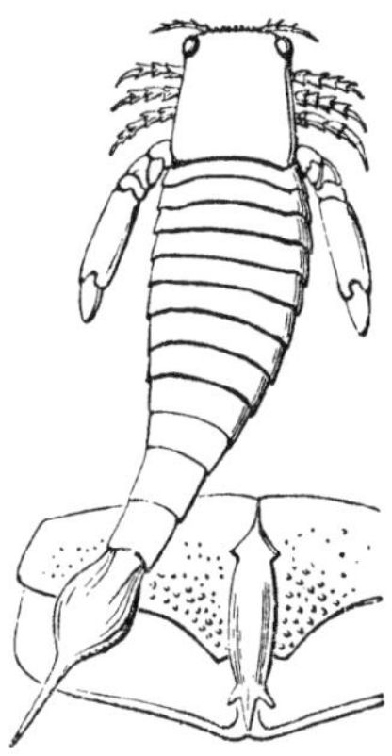 Slimonia acuminata, with thoracic plate.	**Slimonia acuminata,** Salter (*Pterygotus acuminatus*, Mon. Pteryg. pl. 2). The thirteen body-rings, including the long acuminate tail joint, the upper and under sides, antennæ, palpi, metastoma or post-oral plate, the thoracic plate (of male and female), and the swimming feet may be seen in the specimens. There is a large and nearly perfect specimen (b. 161) placed over the Woodwardian Cabinet.	b. 38, b. 39, Lesmahagow, Lanark; b. 142, b. 144, b. 145, b. 146, shew thoracic plates; b. 129, b. 131, shew the metastoma, long and narrow; b. 143, the largest head known; b. 122, head of young specimen; b. 123, b. 124, shew swimming feet and metastoma *in situ;* b. 122, a palpus, very large; b. 126, shews all these and body, Lesmahagow.
G g 2		**Slimonia,** sp. new ? metastoma.	b. 130, Lesmahagow.
		BRACHIOPODA.	
G f 1	p. 253 ?	**Lingula,** like *L. lata*, Sowerby.	b. 10, Benson Knott, Kendal.
G f 1	*L. cornea*, p. 251. 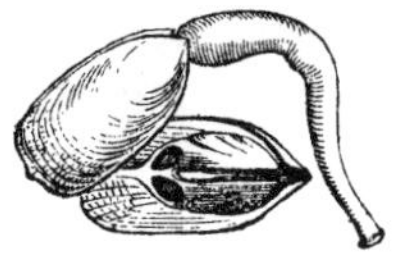	**Lingula minima,** Sow. (Sil. Syst. t. 5, fig. 23, Davids. Sil. Brach. t. 2, figs. 36—44).	b. 957, Woolhope; b. 958, near Ludlow.

Case and Column of Drawers.	Reference to McCoy's Synopsis: and Figures of Genera.	Names and References; Observations, &c.	Numbers and Localities.
G g 2		**Lingula,** sp. like **minima,** Salter (Explan. Sheet 32, Mem. Geol. Surv.).	b. 138, Lesmahagow.
G f G f	p. 251.	**Lingula cornea,** Sow. (Siluria, 2nd ed. pl. 34, fig. 2). (See *L. minima*, p. 179.)	b. 956, near Ludlow; b. 48, Benson Knott, Kendal.
G f 1	p. 253.	**Lingula Lewisii?** Sow. (These specimens are fine, and possibly a large variety of *L. cornea*, Davids. Sil. Brach. t. 3, figs. 1—6.)	b. 14, Benson Knott, Kendal.
G f	*Discina*, p. 191.	**Trematis (Discina) striata** (Davidson, Sil. Brach. t. 6, figs. 1—4).	b. 958, Benson Knott.
G f 1 G f	p. 190.	**Discina rugata,** Sow. (Siluria, 2nd ed. pl. 20, figs. 1, 2, Davidson, Sil. Brach.).	b. 959, Woolhope; b. 960, near Ludlow; b. 961, Potter Fell, Kendal; b. 962, Benson Knott, Kendal; b. 963, Burton and Brockton.
G f 1		**Discina,** sp. like **Forbesii** (perhaps same).	b. 964, Park Lane, Llandeilo.
		Crania (see p. 135).	
	Orbiculoidea, p. 189.	**Crania implicata,** Sow. (Siluria, 2nd ed. pl. 20, fig. 4. Davids. l. c. pl. 8, fig. 13). A very small species well described by Davidson.	Park Lane, Llandeilo.
G f 1	*S. subspuria*, p. 195.	**Spirifer elevatus,** Dalm. (see p. 138), equally common in Gothland, and varying much. *S. bijugosa*, McCoy, from rocks of this age in W. Ireland, is a kindred species.	b. 965, N. end of Potter Fell; b. 967, Benson Knott.
G f 1 G f	*Hemithyris*, p. 204.	**Rhynchonella nucula,** Sow. (see p. 167). The commonest of all Ludlow shells, except its constant companion *Chonetes lata*. Both began in the May Hill sandstone and multiplied when sandy strata and shallow water prevailed.	b. 968, Lambrigg Fell, Kendal; b. 969, Burton and Brockton (and as *Davidsoni*); b. 970, Benson Knott, Kendal; b. 971, Collinfield, Kendal; b. 972, Mortimer's Cross, Aymestry; b. 973, Woolhope; b. 974, near Ludlow.
G f 1	*Hemithyris*, p. 207.	**Rhynchonella Wilsoni,** Sow. (see p. 167).	b. 975, Burton and Brockton.
G f 1	*H. Davidsoni*, p. 200.	**Rhynchonella Wilsoni,** var. **Davidsoni,** McCoy (see p. 141).	b. 977, Burton and Brockton.

Case and Column of Drawers.	Reference to McCoy's Synopsis: and Figures of Genera.	Names and References; Observations, &c.	Numbers and Localities.
G f 2 G f	p. 220.	**Orthis lunata,** Sow. (Siluria, 2nd ed. pl. 20, fig. 11). This very fine-ribbed shell is often mistaken; but if the interior characters of the hinge be studied with the outside, no difficulty ought to be felt in separating it from its companion *O. orbicularis,* a coarser shell with fasciculate striæ. The teeth of the dorsal valve are nearly parallel and the central cardinal process *linear.*	b. 976, Kington, Herefordshire; b. 978, Woolhope; b. 979, Mortimer's Cross, Aymestry; b. 980, Downton Castle, Ludlow; b. 981, Leintwardine; b. 982, Benson Knott, Kendal.
G f 2	*Leptæna,* p. 243.	**Strophomena filosa,** Sow. (see p. 146). More common in Ludlow rocks than elsewhere; at Ludlow it varies exceedingly in the sculpture.	b. 983, Burton and Brockton, near Wenlock.
G f 2	*Leptagonia,* p. 248.	**Strophomena depressa,** Sow. (see p. 147).	b. 984, Burton and Brockton, near Wenlock.
G f G f 2	*Leptæna,* p. 249.	**Chonetes lata,** Von Buch, (regarded by some as the true *C. striatella* of Dalman) but the references in the 3rd ed. of Siluria are so confused and sterile that no use can be made of them. The best fig. is that in Siluria, 2nd ed. pl. 20, fig. 8.	b. 985, near Woolhope; b. 986, Benson Knott, Kendal; b. 987, Lambrigg Fell, Kendal; b. 988, Brockton and Burton; b. 989, Mortimer's Cross, Aymestry; b. 990, near Ludlow; b. 991, Downton Castle.
		LAMELLIBRANCHIATA.	
G f G f 5	Pl. 1 I, figs. 11—15, p. 258.	**Avicula Danbyi,** McCoy. Generally wide oval, a large fine shell. It varies much.	b. 49 to b. 52, Benson Knott, Kendal; b. 52, right valve; b. 51 is a rounder variety; b. 955, left valve.
G f 4	p. 261.	**Pterinea lineata,** Goldf.? (Pet. Germ., t. 119, fig. 6). A beautiful shell, close to *P. retroflexa,* but with long wing, and very fine regular radiating ridges of striæ. (See *P. lineatula.*)	b. 85, Benson Knott, Kendal.
G f G f 4	*P.* var. *naviformis,* p. 263.	**Pterinea retroflexa,** Wahl. sp. (His. Leth. Suec. t. 17, fig. 12). Has convex right valve, var. *naviformis* of Conrad.	b. 83, b. 84, Kendal, type variety; b. 82, right valve, Kendal; b. 19, Kirkby Moor, Kendal.
G f	p. 262.	**Pterinea retroflexa,** var. **erecta,** McCoy.	b. 62, Laverock Lane, Kendal.
	Pterinea lineata, p. 261.	**Pterinea lineatula,** D'Orb. (Siluria, pl. 23, fig. 16). Extremely common in Ludlow rock.	

Case and Column of Drawers.	Reference to McCoy's Synopsis: and Figures of Genera.	Names and References; Observations, &c.	Numbers and Localities.
G f 4	Pl. 1 I, fig. 7, p. 260.	**Pterinea demissa,** Conrad. A variety of *P. retroflexa*, squarer and less falcate at the wing.	b. 18, Benson Knott.
G f 4		**Pterinea demissa,** var. with more point to wing.	b. 17, Benson Knott.
G f 4	Pl. 1 I, fig. 3, fig. 263.	**Pterinea subfalcata,** Conrad. Regular squamæ of growth, and irregular radiating striæ.	b. 992, Benson Knott; b. 22, Pont-ar-y-llechau, Llangadock.
G f 4	*P. Boydi*, p. 259.	**Pterinea Sowerbyi,** McCoy, p. 263. Beautiful variety of it.	b. 87, Brigsteer, Kendal.
G f 4	Pl. 1 I, fig. 4, p. 263.	**Pterinea tenuistriata,** McCoy. Like a small *P. retroflexa*, but with radiating striæ.	b. 21, near Ludlow; b. 20, b. 86, Benson Knott, Kendal.
G f 4	Pl. 1 I, figs. 1, 2, p. 261.	**Pterinea pleuroptera,** Conrad. Has shorter wing than *P. retroflexa*, and some ridges on the posterior line.	b. 993, Benson Knott, Kendal.
G f 5	*M. complanata*, p. 266?	**Modiolopsis planata,** Salter. Flat area, and valves slightly keeled.	b. 72, b. 73, var. Kirkby Moor, Kendal; b. 71, Kendal.
G f 5	*M. solenoides*, p. 269.	**Modiolopsis planata,** var. longer, and pressed in stone (not of Sowerby, which is *Orthonotus*).	b. 92, Benson Knott.
G g		**Modiolopsis,** sp. unnamed.	b. 994, Lesmahagow, Lanark.
G f 5	*Anodontopsis*, p. 271.	**Modiolopsis lævis,** Sow. (Siluria, 2nd ed. pl. 34, fig. 7).	b. 995, Llechclawdd, Myddfai, Llandovery.
G f 5	Pl. 1 K, figs. 14, 15, p. 271.	**Anodontopsis angustifrons,** McCoy. Two varieties, a very broad and a narrower one. Some of them are almost circular.	b. 68, Benson Knott; b. 79, Kirkby Moor, Kendal.
G f 5	Pl. 1 K, figs. 11, 12, 13, p. 271.	**Anodontopsis bulla,** McCoy. These rounded shells are but doubtfully of the genus. They resemble the recent *Kellia*.	b. 997, Kirkby Moor, Kendal.
G f	*Orthonotus*, p. 274.	**Goniophora cymbæformis,** Sow. (Siluria, 2nd ed. pl. 34, fig. 15).	b. 996, Burton and Brockton.
G g 2		**Mytilus mimus,** sp. Very like the living mussel.	b. 141, Lesmahagow.

Case and Column of Drawers.	Reference to McCoy's Synopsis: and Figures of Genera.	Names and References; Observations, &c.	Numbers and Localities.
G f 5	*Anodontopsis*, p. 272. Pl. 1 L, fig. 9.	**Pseudaxinus securiformis,** McCoy. The genus was founded to include thin shells with no teeth, and the aspect of *Trigonia*. They are related to the *Modiolæ*.	b. 74, b. 75, vars. Benson Knott, Kendal.
G f 5	*Sanguinolites*, p. 277. Pl. 1 I, fig. 24.	**Orthonotus decipiens,** McCoy.	b. 69, Benson Knott.
G f	*Sanguinolites*, p. 276. Pl. 1 K, figs. 19, 20.	**Orthonotus anguliferus,** McCoy. A new square species related to *O. solenoides*, Sow. but with epidermal marks like *Mya v.-scripta*.	b. 42, Benson Knott.
G f G f 5	*Leptodomus*, p. 278. 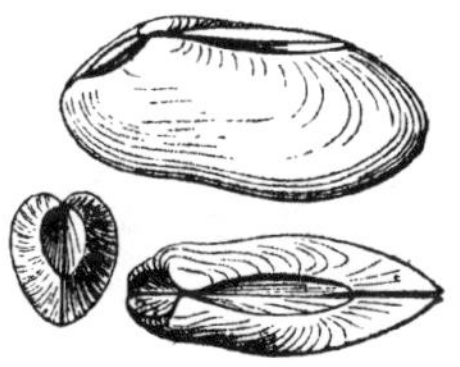	**Orthonotus amygdalinus,** Sow. (Siluria, 2nd ed. pl. 23, fig. 6). A rather rougher variety of this usually smooth shell.	b. 998, near Ludlow; b. 999, Benson Knott; c. 1, Brigsteer, Kendal; c. 2, Cwm Craig ddu, Builth.
G f 5		**Orthonotus amygdalinus,** var. **retusus,** Sow. (Siluria, 2nd ed. pl. 23, fig. 7).	c. 3, near Ludlow.
G f G f 5	*Leptodomus*, p. 278. Pl. 1 L, fig. 11.	**Orthonotus amygdalinus,** Sow. var. **globulosus,** McCoy. A variety of *O. amygdalinus* evidently.	c. 4, Kirkby Moor, Kendal; b. 43, Benson Knott; c. 5, Tenter Fell, Kendal.
G f 5	*Leptodomus*, p. 279. Pl. 1 K, figs. 21—24.	**Orthonotus truncatus,** McCoy. A well-marked shell.	b. 88, b. 89, b. 90, Benson Knott; b. 91, Mortimer's Cross.
G f 5	*Leptodomus*, p. 279.	**Orthonotus undatus,** Sow. (Siluria, 2nd ed. pl. 23, fig. 4), ridges regular.	b. 70, Benson Knott.
G f	*Orth. semisulcatus*, p. 275. Pl. 1 K, fig. 25.	**Orthonotus prora,** Salter (Siluria, 3rd ed. Foss. 61, fig. 4). Easily known by the narrow front.	b. 63, Kirkby Moor, Kendal.
G f 5	*Tellinites?* p. 286. Pl. 1 K, fig. 31.	**Orthonotus affinis,** McCoy. Certainly an *Orthonotus* of the usual shape and ornament.	b. 93, Benson Knott.
	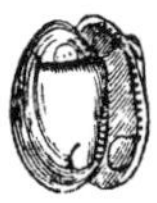	**Cuculella,** McCoy. An useful genus, to include those *Nucula*-like shells, with a strong internal septum on the short posterior side.	
G f G f 4	p. 284.	**Cuculella ovata,** Sow. (Siluria, 2nd ed. pl. 34, fig. 17).	c. 6, Brigsteer, Kendal; b. 64, Benson Knott; c. 7, Derby Arms, Underbarrow.
G f 4	p. 284.	**Cuculella coarctata,** Phill. sp. (Mem. Geol. Surv. Vol. II. Pt. 1, pl. 22, figs. 1—4).	c. 8, Benson Knott, Kendal.

Case and Column of Drawers.	Reference to McCoy's Synopsis: and Figures of Genera.	Names and References; Observations, &c.	Numbers and Localities.
Gf4	p. 283.	**Ctenodonta (Arca) primitiva,** Phill. (Mem. Geol. Surv. Vol. II. Pt. 1, pl. 21, fig. 5).	c. 8*, Benson Knott, Kendal.
Gf4	(*Arca*) Pl. 1 K, figs. 2, 3, p. 283.	**Ctenodonta Edmondiiformis,** McCoy? Surely a *Ctenodonta.*	c. 9, Benson Knott.
Gf4	*Nucula,* p. 285.	**Ctenodonta anglica,** D'Orb. (Siluria, 2nd ed. pl. 23, fig. 10).	c. 10, Brigsteer, Kendal.
		Grammysia, De Vern. (1837). The constrictions along the course of the chief pedal muscles distinguish this from *Orthonotus.* Many fine forms occur in North America.	
Gf4	p. 280.	**Grammysia cingulata,** His. sp. var. **obliqua.**	b. 67, High Thorns, Underbarrow, Kendal.
Gf4	do.	**Grammysia cingulata**? Hisinger (Mem. Geol. Surv. Vol. II. Pt. 1, pl. 17, fig. 3).	c. 11, Benson Knott.
Gf	Pl. 1 K, fig. 29, p. 281.	**Grammysia extrasulcata,** Salter (Mem. Geol. Surv. Vol. II. Pt. 1, pl. 17, fig. 2).	b. 45, Benson Knott, good left valve.
Gf	Pl. 1 K, fig. 28, p. 280.	**Grammysia triangulata,** Salter (Mem. Geol. Surv. Vol. II. Pt. 1, pl. 17, fig. 1).	b. 47, right valves, b. 46, left, Benson Knott.
Gf	Pl. 1 K, figs. 26, 27, p. 281.	**Grammysia? rotundata,** Sow. sp. (*Orthonota,* Siluria, 3rd ed. pl. 23, fig. 5). Scarcely different from *Orthonotus,* the constriction being quite absent.	b. 44, Benson Knott, Kendal.
		PTEROPODA.	
	Pteropods.	An inferior group of Molluscs to ordinary univalves.	
Gf6	p. 287.	**Theca Forbesii,** Sharpe (Quart. Journ. Geol. Soc. Vol. II. p. 314, t. 13, fig. 1).	c. 12, Lambrigg Fell, Kendal.

Case and Column of Drawers.	Reference to McCoy's Synopsis: and Figures of Genera.	Names and References; Observations, &c.	Numbers and Localities.
G f 6	*C. cancellata,* p. 287.	**Conularia Sowerbyi,** Defr. (Siluria, 2nd ed. pl. 25, fig. 10. *C. quadrisulcata,* Sow. Sil. Syst. t. 12, fig. 22).	c. 13, Underbarrow, Kendal; c. 14, Benson Knott, Kendal.
G f G f 6	Pl. 1 L, fig. 24, p. 288.	**Conularia subtilis,** Salter. Appendix, p. vi. Very fine close ridges.	c. 15, Brigsteer, b. 59, Benson Knott.
		[**Cyrtolites lævis,** Sow. Siluria, pl. 25, fig. 9, is very common in these rocks.]	
	Univalves.	**GASTEROPODA.**	
G f 6		**Pleurotomaria,** sp.	c. 16, Kendal.
G f	Pl. 1 K, fig. 45, p. 291.	**Pleurotomaria crenulata,** McCoy.	b. 66, Brigsteer.
G f 6	Pl. 1 L, fig. 19, p. 294.	**Murchisonia torquata,** McCoy.	c. 17, Benson Knott; c. 18, Spital, Kendal.
G f G f 6	Also p. 303, *Holopella cancellata ?*	**Murchisonia articulata,** Sow. (Siluria, 2nd ed. pl. 24, fig. 2).	c. 19, Benson Knott; c. 19*, Lambrigg Fell, Kendal.
G f	*Litorina,* p. 305.	**Cyclonema corallii,** Sow. (Siluria, 2nd ed. pl. 24, fig. 1).	b. 60, Benson Knott.
G f 6	*Litorina,* p. 305.	**Cyclonema octavia,** D'Orb. Sow. (Siluria, 2nd ed. pl. 24, fig. 4. *Turbo carinatus,* Sow. Sil. Syst. t. 5, fig. 28).	c. 20, Burton and Brockton, Woolhope.
G f 6	*N. glaucinoides,* p. 302.	**Naticopsis parva,** Sow. (Siluria, 2nd ed. pl. 25, fig. 1).	c. 21, Beckfoot, Kirkby Lonsdale; c. 22, Benson Knott.
G f 6	p. 303.	**Holopella gregaria,** Sow. (Siluria, 2nd ed. pl. 34, fig. 10 *a*).	c. 23, Underbarrow; c. 24, Beckfoot, Kirkby Lonsdale.

Case and Column of Drawers.	Reference to McCoy's Synopsis: and Figures of Genera.	Names and References; Observations, &c.	Numbers and Localities.
G f 6	p. 304.	**Holopella obsoleta,** Sow. (Siluria, 2nd ed. pl. 34, fig. 11). A fine shell, everywhere characteristic of sandy Ludlow rock.	c. 25, Benson Knott.
G g G g 2		**Platyschisma helicoides,** Salter. A shell very like the *Trochus helicites* of the British Ludlows, but flatter and having a marked subangular band.	b. 140, c. 26, Lesmahagow.
		HETEROPODA (Nucleobranchiata).	
G f	p. 309.	**Bellerophon expansus,** Sow. (Siluria, 2nd ed. pl. 25, fig. 8).	c. 27, near Ludlow.
G f		**Bellerophon Ruthveni,** n. s. Smaller than *B. dilatatus,* and with the band angular, and the whorls angular where the band becomes so. Very common, $1\frac{1}{4}$ inch wide.	b. 61, Benson Knott.
		CEPHALOPODA.	
G f G f 7	*Cycloceras,* p. 319.	**Orthoceras Ibex,** Sow. (Siluria, 2nd ed. pl. 29, fig. 3). Often shews sudden expansion, probably breeding season.	c. 28, Meal Bank, Kendal; b. 58, Benson Knott, good.
G f G f 7	p. 313.	**Orthoceras bullatum,** Sow. (Siluria, 2nd ed. pl. 29, fig. 1). The septa shew vascular markings all over them.	c. 29, Lambrigg Fell, Kendal; c. 30, Radnorshire; c. 31, Mortimer's Cross; c. 32, Burton and Brockton.
G f 7	do.	**Orthoceras angulatum,** Wahl. (Siluria, 2nd ed. pl. 28, fig. 4. See p. 173.)	c. 33, Brigsteer, Kendal; c. 34, Underbarrow.
G f 7		**Orthoceras subannulatum?** Not at all likely.	c. 37, Brigsteer.
G g 2		**Orthoceras Maclareni?** Salter (Siluria, 2nd ed. Foss. 24, p. 126).	b. 137, Lesmahagow.
G f 10 G f 7 G f	p. 317.	**Orthoceras tenuicinctum,** Portl. (Geol. Rep. p. 371, t. 27, fig. 5. Siluria, 2nd ed. Foss. 40, fig. 3).	b. 153, Brigsteer; c. 35, Benson Knott, Kendal; c. 36, Woolhope.

Case and Column of Drawers.	Reference to McCoy's Synopsis: and Figures of Genera.	Names and References; Observations, &c.	Numbers and Localities.
G f 7	Appendix, p. vii.	**Orthoceras torquatum,** Münst. Salter in Appendix, p. vii. McCoy considers this a doubtful species. I do not.	c. 38, Benson Knott.
G f 7	*Cycloceras,* p. 320. Pl. 1 L, fig. 31.	**Orthoceras tenuiannulatum,** McCoy. Rings slender, and striæ equal.	c. 39, Brigsteer, Kendal.
G f 7	*O. laqueatum,* p. 315.	**Orthoceras,** sp. with fine thread-like ridges. (*O. filosum,* young?)	c. 40, Kirkby Moor, Kendal.
G f G f 7	(*Cycloceras*), p. 321.	**Orthoceras tracheale,** Sow. (Siluria, 2nd ed. pl. 34, fig. 6).	c. 41, Kirkby Moor; c. 42, near Ludlow.
G f	p. 315.	**Orthoceras imbricatum,** Wahl. (Siluria, 2nd ed. pl. 29, fig. 7).	c. 43, Kirkby Moor, near Kendal.
G f 7	*O. subundulatum,* p. 317.	**Orthoceras,** new, large; ten times the size of Portlock's shell.	c. 44, High Thorns, Underbarrow.
G f 7	p. 314. Appendix, p. vii.	**Orthoceras dimidiatum,** Sow. (Siluria, 2nd ed. pl. 28, fig. 5). Easily known by the squamate ridges reaching half across only.	c. 45, Brigsteer.
G f 7	Pl. 1 L, fig. 27, p. 313. Appendix, p. vi.	**Orthoceras baculiforme,** Salter, Appendix to Synopsis. The smooth species are difficult, but not unrecognizable. This is oval in section and very long and tapering indeed.	c. 46, Brigsteer (the smaller specimen is coated with tubercular sponge, c. 47).
	Hortolus Ibex, p. 324.	**Orthoceras perelegans,** Salter (Mem. Geol. Surv. Vol. II. Pt. 1, t. 13, figs. 2, 3). McCoy in his attempt to combine the disjointed figures of this species, *Orthoceras Ibex* and *Orthoceras tracheale,* has again confused the irregular curving *Orthoceras? perelegans* with the straight shell *O. tracheale.*	

LUDLOW BONE-BED AND DOWNTON SANDSTONE.

The Bone-bed, indicating a long period of rest, covers the sandy Upper Ludlow rocks over a large area. At Ludlow it is divided into two thin bands not an inch thick. At Norton, near Onibury, it is a foot thick, one mass of fish defences (Onchus), shagreen of some shark-like or *Acanthodian* species, and shells of Brachiopoda—Discina and Lingula—with a few *Pteraspid* fish. The whole appearance is that of a drift in shallow but quiet water, and the fact that the Bone-bed is everywhere covered by the Downton Sandstone rock shows that the deep-sea condition had passed, and the coast-line raised. West of the central Silurian region the Bone-bed is not known; but within that region it ranges from Ludlow to Malvern, thence to Woolhope and Hagley; but fails at Usk, though the fish defences and bones of the same species are found in Red Downton rock of coarse texture.

The Downton Sandstone is the true top of the Silurian system, and contains several Ludlow shells. Over this, the red sediments, evidently accumulated in land-locked seas and estuaries, contain abundance of Pteraspid and Cephalaspid fish; and only a single shell, *Lingula cornea,* and a few crustacea, *Pterygotus Banksii* and *Beyrichia,* rise into the base of the Old Red Sandstone (Ledbury Shales).

(Note. The term *Tilestones* is technically applied to the flaggy beds at the top of the Ludlow series of rocks.)

Case and Column of Drawers.	Reference to McCoy's Synopsis: and Figures of Genera.	Names and References; Observations, &c.	Numbers and Localities.
G g 5		A good slab of the *Micaceous Tilestones.*	b. 121, Storm Hill, Llandeilo.
		[Plant fragments, probably of low *Lycopodiaceæ,* are very common in the Silurian district. There are but few in the collection.]	Hagley, Ludlow, &c.
		Pachytheca sphærica, Hooker (Siluria, 3rd ed. pl. 35, fig. 30. Hooker, Quart. Journ. Geol. Soc. Vol. XVII. p. 162). This very common seed-case or spore of (probably) a water-plant abounds in Downton Sandstone.	Woolhope.
		Actinophyllum or **Spongarium?** or some allied plant half an inch in diameter, is frequent near Ludlow.	Ludlow.
		CRUSTACEA (*Merostomata,* Dana).	
		Pterygotus problematicus, Ag., 6 ft. long.	Ludlow.

Case and Column of Drawers.	Reference to McCoy's Synopsis: and Figures of Genera.	Names and References; Observations, &c.	Numbers and Localities.
		Pterygotus gigas, Salter (Mem. Geol. Surv. Mon. 1, pl. 9), 6 or 7 ft. long.	Kington.
		Pterygotus Banksii, Salter (Siluria, 3rd ed. p. 239). A small species.	Ludlow.
		Hemiaspis, one or two species.	
		CRUSTACEA. Phyllopoda.	
	Pl. 1 E, fig. 2, p. 135.	**Beyrichia Klödeni,** McCoy. Very abundant.	Ludlow and S. Wales.
		Leperditia marginata, Keyserling?	
		CRUSTACEA. Trilobita.	
		Trilobites are excessively rare, and *Echinodermata* and corals quite absent. The inhabitants of a sandy shore and such spoils of neighbouring lands as might be preserved on a shallow coast-line are all we can expect.	
		ANNELIDA.	
	Worm-tracks.	**Crossopodia,** McCoy. The deep trail of a large worm. Annelids were gigantic in Silurian times. One (*Beatricea*) 30 feet in length! and thick as a man's thigh (Logan).	
G g	Pl. 1 D, fig. 14, p. 130.	**Crossopodia lata,** McCoy. The trail of the body is very distinct from that of the cirri or lateral appendages.	b. 157, Storm Hill, Llandeilo.
		BRACHIOPODA.	
		Brachiopods except Lingula are very rare.	

Case and Column of Drawers.	Reference to McCoy's Synopsis: and Figures of Genera.	Names and References; Observations, &c.	Numbers and Localities.
		Lingula cornea, Sowerby (Siluria, 3rd ed. pl. 34, fig. 2).	
		Lingula minima, Sow. (Sil. Syst. t. 5, fig. 23).	Downton Sandstone.
		Discina rugata, Sow. (Siluria, 3rd ed. pl. 20, figs. 1, 2).	Bone-bed.
		Chonetes lata, and **Orthis lunata.**	Pont-ar-y-llechau, Llangadock.
		LAMELLIBRANCHIATA.	
G g 6	*M. solenoides*, p. 269.	**Modiolopsis planata,** Salter, n. sp.	b. 104, Storm Hill, Llandeilo.
G g		**Modiolopsis,** sp. More convex than *M. complanata.*	b. 160, Downton, with *Platyschisma helicites.*
G g 6	*M. complanata*, p. 266 ?	**Modiolopsis lævis,** Sow. (Siluria, 2nd ed. pl. 34, fig. 7).	b. 108, Storm Hill, Llandeilo.
G g	p. 268.	**Modiolopsis platyphyllus,** Salter (Mem. Geol. Surv. Vol. II. Pt. 1, pl. 20, figs. 13, 14).	b. 159, Storm Hill, Llandeilo, figured.
G g 6	p. 262.	**Pterinea retroflexa,** Hisinger (Siluria, 2nd ed. pl. 9, fig. 26; pl. 23, fig. 17).	c. 48, Horeb Chapel.
G g 6	Pl. 1 I, fig. 4, p. 263.	**Pterinea tenuistriata,** McCoy. (See p. 182.)	b. 117, Pont-ar-y-llechau, Llangadock.
G g 6	Pl. 1 I, figs. 19, 20, p. 261.	**Pterinea megaloba,** McCoy. A very convex species, more like an inflated *Modiola.*	b. 105, b. 106, b. 107, Storm Hill, Llandeilo.
G g 6	Pl. 1 I, fig. 7, p. 260.	**Pterinea demissa,** Conrad. (See p. 182.)	b. 114, Pont-ar-y-llechau.
G g 6	Pl. 1 K, fig. 1, p. 283.	**Ctenodonta subæqualis,** McCoy, sp. (Siluria, 2nd ed. pl. 10, figs. 7, 8).	c. 49, Storm Hill, Llandeilo; b. 115, Llechclawdd, Myddfai.
G g 6	*Anodontopsis quadratus*, Pl. 1 K, fig. 10, p. 272.	**Palæarca diagona,** Salter, n. sp. Almost a rhomb in shape.	b. 110, b. 111, Storm Hill, Llandeilo.

Case and Column of Drawers.	Reference to McCoy's Synopsis: and Figures of Genera.	Names and References; Observations, &c.	Numbers and Localities.
G g	*Dolabra.* Pl. 1 K, fig. 30, p. 270.	**Palæarca (Dolabra) obtusa,** McCoy.	b. 158, figured, Storm Hill, Llandeilo.
G g 6	Pl. 1 L, fig. 10, p. 269. 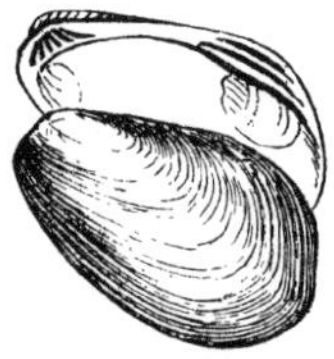	**Palæarca? elliptica,** McCoy.	b. 116, Storm Hill.
G g 6		**Cuculella ovata,** Sow. sp. (Siluria, 2nd ed. pl. 34, fig. 17).	c. 50, Llechclawdd, Myddfai, Llandeilo; c. 51, Storm Hill, Llandeilo; c. 52, Horeb Chapel.
G g 6	p. 284.	**Cuculella antiqua,** Sow. sp. (Siluria, 2nd ed. pl. 34, fig. 16).	c. 53, Storm Hill, Llandeilo.
G g 6	*Sanguinolites,* Pl. 1 L, fig. 24, p. 277.	**Orthonotus decipiens,** McCoy.	b. 112, Llechclawdd, Myddfai.
G g 6	Pl. 1 K, fig. 28, p. 280.	**Grammysia triangulata,** Salter (Siluria, 2nd ed. Foss. 60, fig. 2).	c. 54, Storm Hill.
G g	*Orthonotus,* p. 274.	**Goniophora cymbæformis,** Sow. (Siluria, 2nd ed. pl. 23, fig. 2; pl. 34, fig. 15).	c. 55, Horeb Chapel.
		GASTEROPODA.	
G g 6	Pl. 1 L, fig. 19, p. 294.	**Murchisonia (Hormotoma) torquata,** McCoy. A small species, a good deal like *M. articulata,* Sow., and allied to it.	b. 118, Storm Hill; c. 56, Pont-ar-y-llechau, near Llangadock.
G g 6	p. 303.	**Holopella gregaria,** Sow. (Siluria, 2nd ed. pl. 34, fig. 10 a).	c. 57, Horeb Chapel, Llandovery.
G g 6	p. 304.	**Holopella obsoleta,** Sow. (Siluria, 2nd ed. pl. 34, fig. 11).	c. 58, Pont-ar-y-llechau, Llangadoc; c. 59, Llechclawdd, Myddfai; c. 60, Horeb Chapel.
G g 6	p. 303.	**Holopella conica,** Sow. (Siluria, 2nd ed. pl. 34, fig. 10; pl. 35, fig. 26).	b. 109, Storm Hill, Llandeilo.
G g G g 6	*Turbo?* p. 296.	**Platyschisma Williamsi,** Sow. (Siluria, 2nd ed. pl. 34, fig. 14). A large and more *Paludina*-like shell than the next and rarer.	b. 113, Llandeilo; c. 61, Horeb Chapel, Llandovery.

Case and Column of Drawers.	Reference to McCoy's Synopsis: and Figures of Genera.	Names and References; Observations, &c.	Numbers and Localities.
G g G g 6	*Trochus?* p. 297.	**Platyschisma helicoides,** Sow. (Siluria, 2nd ed. pl. 34, fig. 12). A shell just like a snail, the surface very smooth. It seems everywhere characteristic of these topmost layers of the Silurian.	c. 62, c. 64, Storm Hill; b. 160, Downton, Ludlow.
		HETEROPODA.	
G g 6	p. 311.	**Bellerophon trilobatus,** Sow. (Siluria, 2nd ed. pl. 34, fig. 9).	c. 63, Storm Hill, Llandeilo.
G g 6	p. 309.	**Bellerophon expansus,** Sow. (Siluria, 2nd ed. pl. 25, fig. 8. (*B. globatus* is the young.)	c. 64, Horeb Chapel, Llandovery.
G g 6	p. 310.	**Bellerophon Murchisoni,** D'Orbigny (Siluria, 2nd ed. pl. 34, fig. 19. *B. striatus,* Sow. Sil. Syst.).	c. 65, Horeb Chapel.
G g	p. 309.	**Bellerophon carinatus,** Sow. (Siluria, 2nd ed. pl. 34, fig. 8).	c. 66, Storm Hill.
		CEPHALOPODA.	
G g 6	*Cycloceras,* p. 321.	**Orthoceras tracheale,** Sow. (Siluria, 2nd ed. pl. 34, fig. 6).	c. 67, Storm Hill, Llandeilo.
G g 6	p. 313.	**Orthoceras bullatum,** Sow. (Siluria, 2nd ed. pl. 29, fig. 1).	c. 68, Storm Hill, Llandeilo.
G g	*Orthoceras,* p. 316.	**Tretoceras semipartitum,** Sow. (Siluria, 2nd ed. pl. 34, fig. 5).	c. 69, Horeb Chapel.
		PISCES.	
G g		**Onchus tenuistriatus,** Ag. (Siluria, 2nd ed. pl. 35, figs. 15, 17).	c. 70, Woolhope, N. end (J. W. Salter, 1867).
G g		Fish defences and coprolites, Siluria, 2nd ed. pl. 35.	c. 71, Ludford Lane, Ludlow [Mr Lightbody, 1866].
		Pteraspis Banksii, Huxley and Salter (Siluria, 3rd ed. Foss. 68, fig. 2).	

Ledbury Shales.

Ledbury Shales (Salter); part of "Tilestone" of Sir R. Murchison. These contain Cephalaspid fish. Plants and one or two Ludlow shells pass up into them, but the fish and crustacea are distinct from those of the Ludlow rocks.

Having been generally included in the Silurian rocks, they are placed here as beds of passage between Silurian and Devonian.

Case and Column of Drawers.	Reference to McCoy's Synopsis: and Figures of Genera.	Names and References; Observations, &c.	Numbers and Localities.
		PLANTÆ[1].	
G 9		**Fragments of Plants—Lycopodiaceæ?**	c. 84, Ludlow Tunnel.
G 9 7		**Plant Remains.**	c. 85, Ludlow.
G 9 7	*Spongarium?* p. 42.	**Actinophyllum plicatum?** Phill. (Mem. Geol. Surv. Vol. II. pt. 1, p. 386, pl. 30, fig. 4).	c. 86, Ludlow Tunnel.
		CRUSTACEA[1]. **Phyllopoda.**	
		Leperditia marginata, Keyserling?	c. 87, Ludlow Tunnel.
G 9			
G 9 7		**Leperditia marginata.**	c. 88, Ludlow Tunnel.
G 9 7		**Leperditia,** sp.	c. 89, Ludlow Tunnel.
G 9 7	Pl. 1 E, fig. 2, p. 135.	**Beyrichia Klödeni,** McCoy.	c. 90, Ludlow Tunnel.
G 9 7		**Fragments of Crustacea.**	c. 91, Ludlow Tunnel.
		CRUSTACEA. Merostomata.	
G 9		**Eurypterus pygmæus,** Salter (Siluria, 3rd ed. Foss. 67, fig. 1, p. 239).	c. 92, Ledbury.
G 9 G 9 7		**Pterygotus Ludensis,** Salter (Mem. Geol. Surv. Mon. 1, pl. 14, figs. 1—13; Siluria, 3rd ed. Foss. 22, fig. 1, p. 140).	c. 93, c. 98, Ludlow Tunnel.
		BRACHIOPODA[1].	
G 9 G 9 7	p. 251.	**Lingula cornea,** Sow. (Siluria, 3rd ed. pl. 34, fig. 2).	c. 94, Ludlow Tunnel.
		PISCES[1].	
G 9 G 9 7		**Cephalaspid Fish.**	c. 95, c. 96, Ludlow Tunnel.
G 9 G 9 7		**Fish spines,** *Onchus.*	c. 97, c. 99, Ludlow Tunnel.

[1] These specimens were presented by Mr. R. Lightbody.

INDEX.

25—2

CAMBRIDGE: PRINTED BY C. J. CLAY, M.A. AT THE UNIVERSITY PRESS.

Printed in the United States
By Bookmasters